ANNALS OF
THE NEW YORK ACADEMY
OF SCIENCES

Volume 979

EDITORIAL STAFF

Executive Editor
BARBARA M. GOLDMAN

Managing Editor
JUSTINE CULLINAN

The New York Academy of Sciences
2 East 63rd Street
New York, New York 10021

THE NEW YORK ACADEMY OF SCIENCES
(Founded in 1817)

BOARD OF GOVERNORS, September 2002 — September 2003

TORSTEN N. WIESEL, *Chairman of the Board*
JOHN T. MORGAN, *Treasurer*

Honorary Life Governors
WILLIAM T. GOLDEN   JOSHUA LEDERBERG

Governors

ELEANOR BAUM   KAREN E. BURKE
LAWRENCE B. BUTTENWIESER   PRAVEEN CHAUDHARI

BRIAN FERGUSON	GERALD FISCHBACH	JOHN H. GIBBONS
MICHAEL GOLDEN	RONALD L. GRAHAM	MARNIE IMHOFF
JACQUELINE LEO	BRUCE McEWEN	PAUL MARKS
RONAY MENSCHEL	JOHN F. NIBLACK	SANDRA PANEM
PETER RINGROSE	JOHN J. ROCHE	LEE G. VANCE
	DEBORAH WILEY	

HELENE L. KAPLAN, *Counsel* [ex officio]

THE LYMPHATIC CONTINUUM
Lymphatic Biology and Disease

ANNALS OF THE NEW YORK ACADEMY OF SCIENCES
Volume 979

THE LYMPHATIC CONTINUUM
Lymphatic Biology and Disease

Edited by Stanley G. Rockson

The New York Academy of Sciences
New York, New York
2002

Copyright © 2002 by the New York Academy of Sciences. All rights reserved. Under the provisions of the United States Copyright Act of 1976, individual readers of the Annals are permitted to make fair use of the material in them for teaching or research. Permission is granted to quote from the Annals provided that the customary acknowledgment is made of the source. Material in the Annals may be republished only by permission of the Academy. Address inquiries to the Permissions Department (editorial@nyas.org) at the New York Academy of Sciences.

Copying fees: *For each copy of an article made beyond the free copying permitted under Section 107 or 108 of the 1976 Copyright Act, a fee should be paid through the Copyright Clearance Center, Inc., 222 Rosewood Drive, Danvers, MA 01923 (www.copyright.com).*

⊚ The paper used in this publication meets the minimum requirements of the American National Standard for Information Sciences—Permanence of Paper for Printed Library Materials, ANSI Z39.48-1984.

Library of Congress Cataloging-in-Publication Data

The lymphatic continuum: lymphatic biology and disease / edited by Stanley G. Rockson
   p.; cm. — (Annals of the New York Academy of Sciences; v. 979)
Includes bibliographical references and index.
  ISBN 1-57331-414-5 (cloth : alk. paper) — ISBN 1-57331-415-3 (paper : alk. paper)
   1. Lymphatics—Physiology—Congresses. 2. Lymphatics—Pathophysiology—Congresses.
 [DNLM: 1. Lymphatic System—Congresses. 2. Lymphatic Diseases—Congresses.
WH 700 L9846 2002] I. Rockson, Stanley G. II. Series.
  Q11.N5 vol. 979
  [QP115]
  500 s—dc21
  [612.4   2002153061

GYAT / BMP
Printed in the United States of America
ISBN 1-57331-414-5 (cloth)
ISBN 1-57331-415-3 (paper)
ISSN 0077-8923

ANNALS OF THE NEW YORK ACADEMY OF SCIENCES

Volume 979
December 2002

THE LYMPHATIC CONTINUUM
Lymphatic Biology and Disease

Editor
STANLEY G. ROCKSON

Conference Organizers
STANLEY G. ROCKSON AND DAVID ZAWIEJA

This volume is the result of a conference entitled **The Lymphatic Continuum**, sponsored by the Lymphatic Research Foundation and held on May 3–4, 2002 in Bethesda, Maryland.

CONTENTS

From the Lymphatic Research Foundation. *By* WENDY CHAITE ix

Introduction

The Lymphatic Continuum: The Past, Present, and Exciting Future of Lymphatic Research. *By* STANLEY G. ROCKSON 1

Part 1. New Tools for Lymphatic Investigation

The Role of the National Center for Research Resources at the National Institutes of Health: Infrastructure to Forge a New Road for Lymphatic Biology and Therapeutics. *By* ANTHONY R. HAYWARD 5

An Overview of the Pathology and Approaches to Tissue Engineering. *By* ERIN R. OCHOA AND JOSEPH P. VACANTI 10

Tissue Engineering of the Lymphatic System. *By* LAURA E. NIKLASON, JENNIFER KOH, AND AMY SOLAN 27

New Tools for Lymphatic Investigation: Panel Discussion 35

Part 2. Biological Principles

Research Perspectives in Inherited Lymphatic Disease. *By* ROBERT E. FERRELL 39

Contractility Patterns of Normal and Pathologically Changed Human Lymphatics. *By* WALDEMAR L. OLSZEWSKI. 52

Preclinical Models of Lymphatic Disease: The Potential for Growth Factor and Gene Therapy. *By* STANLEY G. ROCKSON 64

Biological Principles: Panel Discussion 76

Part 3. Vascular Development

Placental Growth Factor (PlGF) and Its Receptor Flt-1 (VEGFR-1): Novel Therapeutic Targets for Angiogenic Disorders. *By* AERNOUT LUTTUN, MARC TJWA, AND PETER CARMELIET 80

Insights into the Molecular Pathogenesis and Targeted Treatment of Lymphedema. *By* ANNE SAARISTO, MARIKA J. KARKKAINEN, AND KARI ALITALO 94

Lymphangiogenesis: New Mechanisms. *By* LYNN CHANG, ARJA KAIPAINEN, AND JUDAH FOLKMAN 111

Lymphatic Vessel Activation in Cancer. *By* MELANIE CASSELLA AND MIHAELA SKOBE 120

Part 4. Aspects of Lymphatic Biology and Disease

The Pathogenesis of Filarial Lymphedema: Is it the Worm or Is It the Host? *By* PATRICK J. LAMMIE, KAREN T.CUENCO, AND GEORGE A. PUNKOSDY . . 131

The Molecular Control of Adipogenesis, with Special Reference to Lymphatic Pathology. *By* EVAN D. ROSEN 143

A Stepwise Model of the Development of Lymphatic Vasculature. *By* GUILLERMO OLIVER AND NATASHA HARVEY 159

De Novo Lymph Node Formation in Chronic Inflammation of the Human Leg. *By* WALDEMAR L. OLSZEWSKI 166

Physiologic Aspects of Lymphatic Contractile Function: Current Perspectives. *By* ANATOLIY A. GASHEV. 178

Aspects of Lymphatic Biology and Disease: Panel Discussion 188

Part 5. New Horizons

The Role of Interstitial Stress in Lymphatic Function and Lymphangiogenesis. *By* MELODY A. SWARTZ AND KENDRICK C. BOARDMAN, JR. 197

Proteomic Technologies to Study Diseases of the Lymphatic Vascular System. *By* LEE V. LEAK, EMANUEL F. PETRICOIN, III, MICHAEL JONES, CLOUD P. PAWELETZ, ALI M. ARDEKANI, VINCENT A. FUSARO, SALLY ROSS, AND LANCE A. LIOTTA 211

New Horizons: Panel Discussion 229

Index of Contributors 235

Assistance was received from:

- BIO COMPRESSION SYSTEMS, INC.
- CIRCAID
- GENOMICS COLLABORATIVE, INC.
- GREENBERG-MAY FOUNDATION
- LENORE R. BLAND & SYDNEY F. BLAND CHARITABLE FOUNDATION
- MERCK, INC.
- NATIONAL INSTITUTES OF HEALTH
  — NATIONAL HEART, LUNG, AND BLOOD INSTITUTE
  — NATIONAL CENTER FOR RESEARCH RESOURCES
  — NATIONAL CANCER INSTITUTE
  — NATIONAL INSTITUTE OF ALLERGY AND INFECTIOUS DISEASES
  — NATIONAL INSTITUTE OF CHILD HEALTH AND DEVELOPMENT
  — OFFICE OF RARE DISEASES
- NATIONAL LYMPHEDEMA NETWORK
- NEW YORK ACADEMY OF SCIENCES
- PFIZER, INC.
- REGENERON PHARMACEUTICALS, INC.
- THE SUSAN G. KOMEN BREAST CANCER FOUNDATION, INC.
- WOLLOWICK FAMILY FOUNDATION
- YABLICK CHARITIES, INC.

> The New York Academy of Sciences believes it has a responsibility to provide an open forum for discussion of scientific questions. The positions taken by the participants in the reported conferences are their own and not necessarily those of the Academy. The Academy has no intent to influence legislation by providing such forums.

From the Lymphatic Research Foundation

WENDY CHAITE

Lymphatic Research Foundation, Roslyn, New York 11576, USA

As Founder and President of the Lymphatic Research Foundation (LRF), I am grateful to the New York Academy of Sciences for publishing the proceedings of the conference entitled **The Lymphatic Continuum**. Founded in July 1998, the LRF is a 501(c)(3) not-for-profit, tax-exempt organization whose mission is to promote and support lymphatic research that will lead to therapeutic advance and, ultimately, a cure for lymphatic disease, lymphedema, and related disorders.

The LRF's program goals include fostering related medical and scientific disciplines to participate in lymphatic research; increasing public and private funding and support for lymphatic research by government, industry, foundations, and individuals; stimulating collaboration and the exchange of information and resources within the scientific and medical community; providing research grants and awards; and promoting a national tissue bank and patient registry.

Not only are there exciting opportunities in the scientific arena for lymphatic research, but also national support is emerging from government, industry, private corporations, and individuals. As a result of the LRF's advocacy efforts, significant groundwork has been laid to help foster and support this important field of research and medicine.

Recent advancements include:

- Establishment of a Gordon Research Conference series entitled *Molecular Mechanisms of Lymphatic Function and Disease* to be held biennially, commencing in 2004.

- Launching of a high-quality, international peer-reviewed journal, *Lymphatic Research and Biology*; for manuscript submission and subscription information, visit www.liebertpub.com/lrb

Address for correspondence: Wendy Chaite, Esq., Lymphatic Research Foundation, 39 Pool Drive, Roslyn, NY 11576. Voice: 516-625-9675; fax: 516-625-9410.
lrf@lymphaticresearch.org
www.lymphaticresearch.org

- Creation of a Trans-NIH Coordinating Committee for Lymphatic Research and Disease made up of NIH staff from various Institutes.
- Pronouncement of a NHLBI Request for Applications (RFA) addressing "Functional Heterogeneity of the Peripheral, Pulmonary, and Lymphatic Vessels." Letter of Intent receipt date: 1/14/03. Application receipt date: 2/15/03 (http://grants1.nih.gov/grants/guide/rfa-files/RFA-HL-03-004.html).
- Issuance of a NIH Program Announcement calling for grant applications on the pathogenesis and treatment of lymphedema (www.grants.nih.gov/grants/guide/pa-files/PA-01-035.html).
- Appearance of Senate Appropriations Committee report language strongly urging the NIH to stimulate and support intramural and extramural programs for basic and translational research relating to lymphatic disease and to examine whether experts on lymphatic research are adequately represented on CSR peer review panels and relevant Institute study sections.
- Two international lymphatic research and disease "think tank" conferences held at and with the support of the National Institutes of Health (NIH).
- Establishment of LRF's Lymphatic Research Grants and Awards Program to encourage young investigators to conduct lymphatic research and to recognize outstanding leaders who are helping to advance the field of research.

The lymphatic system is an exquisitely important, functional component of the human body, yet there has been, both historically and currently, an inexplicable, widespread neglect of this organ system. The LRF is committed to promoting and supporting basic and translational research so that greater scientific advancements can be made. The LRF is also committed to advancing lymphatic studies in medical curricula throughout the United States. Our growth in understanding the lymphatic system, through research, can help bring forth advancements in clinical care for many diseases.

If you are interested in hosting a member of the LRF's Scientific/Medical Advisory Board or Professional Outreach Committee at your institution, to provide an in-service laboratory or grand rounds presentation, please contact Wendy Chaite at wchaite@lymphaticresearch.org.

The Lymphatic Continuum
The Past, Present, and Exciting Future of Lymphatic Research

STANLEY G. ROCKSON

Stanford University School of Medicine, Stanford, California 94305, USA

A conference on lymphatic research and biology, entitled *The Lymphatic Continuum*, was convened at the Natcher Center of the National Institutes of Health in Bethesda, Maryland on May 3, 2002. The conference was designed to create an interdisciplinary forum to unite many of the world's leading biomedical investigators, with the intent to discuss current and anticipated developments in lymphatic research. A major subsidiary goal of this conference was to stimulate interest among professionals who have not traditionally studied current lymphatics, yet whose present research and interest directly "interdigitate" with the study of lymphatic disease. It was anticipated that clear directions for future research would emerge that would move this field ahead in the next five to ten years.

The conference was intended to address not only the crying need for intensified research in the field of lymphatic structure, function and disease, but also the interface between lymphatic research and multiple expressions of human disease, such as cancer, infection, metabolism, wound healing and fibrosis, immune disorders, and vascular and developmental biology, among others.

Historically, the birth of lymphatic research occurred in 1627, when a prominent Milanese anatomist, Gasparo Aselli, first recognized and described the milky opalescence of the chylomicron-laden visceral lymphatics of the dog intestine (FIG. 1).[1] Anatomic investigations of the lymphatic circulatory system advanced slowly over the centuries, through the pioneering work of such investigators as Pecquet, Bartholinus, Rudbeck, Sappey, and others. It is exciting to contemplate, in the era of ultrastructural anatomic definition, that we have progressed to the point where we can recognize the

---

Address for correspondence: Stanley G. Rockson, M.D., Division of Cardiovascular Medicine, Falk Cardiovascular Research Center, Stanford University School of Medicine, Stanford, CA 94305. Voice: 650-725-7571; fax: 650-725-1599.
srockson@cvmed.stanford.edu

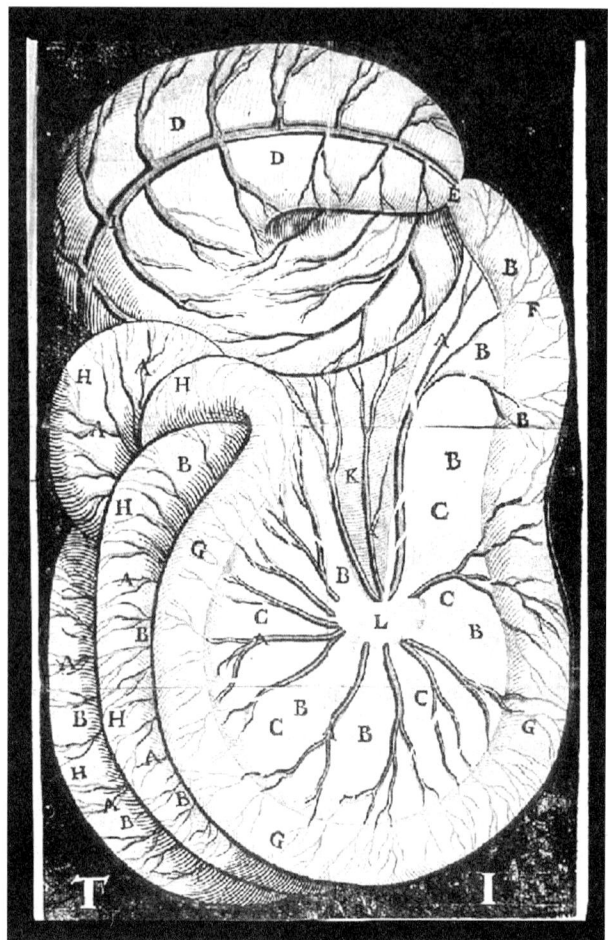

FIGURE 1. Vessels lymphatics of canine intetestine as first described by Gasparo Aselli.

fine reticular filaments that, in large measure, govern the functional relationship between the lymphatic capillary endothelial cell and its interstitial environment (FIG. 2).[2]

A similar exponential trajectory can be utilized to describe our comprehension of the mechanisms that contribute to the expression of human lymphatic disease. In 1892, Milroy first published his description of heredity edema within multiple generations of a single family.[3] Primary, congenital lymphedema is a common disorder of autosomal dominant transmission that eponymously bears his name.[4] Now, scarcely a century after Milroy's initial clinical description, the mechanism of the genetic defect has been linked in numerous

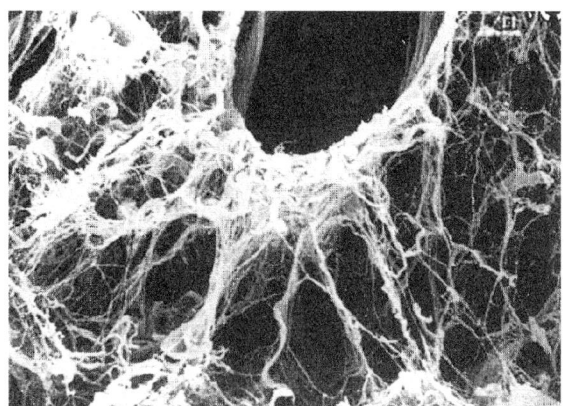

FIGURE 2. Fine reticular filaments govern the functional relationship between the lymphatic capillary endothelial cell and its interstitial environment.

families to missense mutations in the flt4 locus.[5,6] This locus, corresponding to the endothelial receptor (VEGF3) for a lymphangiogenic mitogen (VEGF-C), suggests that Milroy's disease represents a heritable defect of the normal developmental process of lymphangiogenesis.

These, and other, similar, observations have begun to open the door to a new, much anticipated era of molecular therapeutics for diseases of lymphatic structure and function. It has been the goal of the conference and of this volume to explore the exponential growth in technology that can be applied to questions in lymphatic research and to underscore the advances already achieved in our comprehension of the molecular regulatory processes that govern the normal development and function of the lymphatic system.

This volume contains the presentations made at the conference, along with the substantive discussion of the papers that ensued. Many investigative disciplines were represented, including genetics, biochemistry and biophysics, cellular, molecular and developmental biology, physiology, pharmacology, and anatomy. This volume reflects the state-of-the-art approaches to vasculogenesis and angiogenesis, signaling mechanisms, contractile physiology of the lymphatic circulation, and molecular genetics, among others. Avenues to future research are represented by presentations on such topics as tissue engineering, proteomics, and adipogenesis.

It is the hope of the conference organizers that the proceedings of this important conference will serve as a catalyst for intensified investigative efforts in the field of lymphatic research. Ideally, both collaborative efforts and educational imperatives in basic and clinical lymphatic science will be fostered. Indeed, a companion interactive, collaborative workshop to *The Lymphatic Continuum* was held on May 4, 2002. Utilizing the content of this symposium as an intellectual springboard, a select group of investigators and thought leaders were gathered to discuss the imperatives for lymphatic current and future lymphatic research. The proceedings of that workshop, which represent an indispensable adjunct to this volume, are published in the inaugural issue of *Lymphatic Research and Biology,* a new, quarterly, peer-reviewed journal dedicated to the broad, emerging discipline of lymphatic investigation.[7] Additional information about this journal can be obtained from the publisher (http://www.liebertpub.com/LRB).

Certainly, a symposium of this magnitude cannot take shape without the guidance and assistance of many. I am indebted to the following institutes at the NIH for their support: the National Heart, Lung and Blood Institute (NHLBI); the National Center for Research Resources (NCRR); National Cancer Institute (NCI); and the National Institute of Allergy and Infectious Diseases (NIAID). This conference owes much of its existence to my co-chair, Dr. David Zawieja, whose assistance is gratefully acknowledged. Finally, I would like to thank, profoundly, the Founder and President of the Lymphatic Research Foundation, Wendy Chaite, whose vision, energy, and inspiration helps to guide much of the renaissance in lymphatic research.

REFERENCES

1. KANTER, M.A. 1987. The lymphatic system: an historical perspective. Plast. Reconstr. Surg. **79:** 131–139.
2. LEAK, L. & J. BURKE. 1968. Ultrastructural studies on the lymphatic anchoring filaments. J. Cell Biol. 36:129–149.
3. MILROY, W. 1892. An undescribed variety of hereditary oedema. N.Y. Med. J. **56:** 505–508.
4. MILROY, W. 1928. Chronic hereditary edema: Milroy's disease. JAMA **91:** 1172–1174.
5. FERRELL, R.E., K.L. LEVINSON, J.H. ESMAN, *et al.* 1998. Hereditary lymphedema: evidence for linkage and genetic heterogeneity. Hum. Mol. Genet. **7:** 2073–2078.
6. KARKKAINEN, M.J., R.E. FERRELL, E.C. LAWRENCE, *et al.* 2000. Missense mutations interfere with VEGFR-3 signalling in primary lymphoedema. Nat. Genet. **25:** 153–159.
7. ROCKSON, S.G., Ed. 2002. Proceedings of the Conference on Lymphatic Research and Biology. Lymphatic Research and Biology **1:** in press.

The Role of the National Center for Research Resources at the National Institutes of Health

Infrastructure to Forge a New Road for Lymphatic Biology and Therapeutics

ANTHONY R. HAYWARD

Divison of Clinical Research, National Center for Research Resources, National Institutes of Health, Bethesda, Maryland 20892, USA

> ABSTRACT: Lymphatic research has infrastructure needs ranging from the nursing support provided by General Clinical Research Centers to training grants for future clinician investigators. Both have high priority in the activities currently funded by the Division of Clinical Research at NCRR. Further into the future, the therapeutic development networks and embryonic stem cells resources that are currently being developed should seem equally to have been essential resources.
>
> KEYWORDS: lymphatic diseases; research careers; NIH infrastructure; Florence Sabin

Lymphatic diseases, particularly the primary forms of lymphedema, are among the many disorders that are better understood as a result of clinically based research. Advances in the understanding of these relatively rare disorders have come from a better comprehension of the mechanisms of lymphatic development and the genes that contribute to the process. The first step in this process requires the identification of patients and their careful description, that is, phenotyping. The subsequent steps, the identification of chromosomal locations and then genes (through linkage analysis and positional cloning), are increasingly familiar to the readers of medical journals. The

---

Address for correspondence: Anthony R. Hayward, M.D., Ph.D. Director, Division of Clinical Research, NCRR, NIH, 6705 Rockledge Dr., Suite 6030, Bethesda, MD 20892.
haywarda@ncrr.nih.gov

FIGURE 1. Dr. Florence Sabin.

steps themselves are all components of the mission of the Division of Clinical Research at the National Center for Research Resources (NCRR).

NCRR AND INFRASTRUCTURE SUPPORT

In 1960, Public Law 86-798 amended the Public Health Service Act to authorize grants-in-aid to universities, hospitals, laboratories, and other public and nonprofit institutions to strengthen their programs of research and research training in sciences related to health. Grants of this type are the basis for NCRR activities, and they exemplify the support that the National Institutes of Health gives for research resources, as distinct from research on specific disorders. Infrastructure support of this type has been particularly important for the development of research employing human subjects. Milestones in the evolution of the Division of Clinical Research include the creation of the Division of Research Facilities and Resources in 1962, and the merger of the Division of Research Resources and the Division of Research Services to form the NCRR in 1990. In 1995, the present Director of the NCRR, Dr. Vaitukaitis, reorganized the original extramural programs into the following divisions: Biomedical Technology, Clinical Research, Comparative Medicine, and Research Infrastructure. Aside from contributions for the support of specialist conferences, the Division of Clinical Research at the NCRR contributes to lymphatic research through the provision of General Clinical Research Centers (GCRCs) and through training physicians to become clinical researchers.

The network of approximately 80 GCRCs, located primarily at academic medical centers throughout the United States, provides clinical and basic investigators with a specialized environment in which inpatient and outpatient studies can be conducted safely. GCRC resources include research nurses, dietitians, biostatisticians, technicians, and research subject advocates who support patients and investigators day-to-day throughout the research process. These are the resources that have attracted established academic clinician investigators to GCRCs, so that GCRCs may serve as sites for their clinical research programs. The presence of experienced and independently funded investigators within the GCRC has, in turn, made them the focus for clinical training at the institutions where they are found. It is the opportunity for mentoring, and being mentored, that provides much of the excitement in a research career. The experiences of Dr. Florence Sabin, one of the pioneers of lymphatic research, are a fitting example of the important role that mentorship plays in medical advances.

Dr. Sabin, born in Central City, Colorado in 1871, was the first woman to graduate from Johns Hopkins Medical School. Her clinical mentors included Osler. Like Osler, she was attracted to an academic career. Sabin's many "firsts" included an appointment (in 1917) as full professor at a medical college and her election (in 1924) as the first woman president of the American Association of Anatomists. Also, she was the first lifetime woman member of the National Academy of Sciences. Her monograph *The Origin and Development of the Lymphatic System*, published in 1913 by the Johns Hopkins Press,[1] identified the earliest lymph vessels as capillary offshoots of the endothelium of veins. This was an important paradigm shift that ran counter to many of the principal textbooks of her day.

Dr. Sabin, widely identified as a role model (for example, http://www.britannica.com/women/articles/Sabin_Florence_Rena.html), herself published a biography of one of her mentors in anatomy at Johns Hopkins: *Franklin Paine Mall: The Story of a Mind*. It is an appreciation of the importance of mentoring in the training of future clinical researchers that has stimulated the NCRR to expand its funding for K24 and K23 awards and to initiate a K12 program.

NCRR AND TRAINING SUPPORT

Midcareer Investigator (K24) Awards for Patient-Oriented Research from the NCRR support clinicians with a strong training record so they have time protected for patient-oriented research and the mentoring of trainee clinical investigators. Awardees typically are outstanding clinical scientists with a strong record for attracting research support and who are actively engaged in patient-oriented research. The level of grant support allows for up to 50% of their time to be spent in training-related and research activities. National Institutes of Health guidelines require applicants to demonstrate the need for a period of intensive research focus as a means of enhancing their clinical research careers. In addition, applicants must be committed to mentoring the next generation of patient-oriented researchers.

Support for trainees is also a high priority for the NCRR's Division of Clinical Research. Since 1999, K23 awards in Clinical Research have provided up to five years of support for trainees—typically starting in the first or second year following subspecialty training. For these individual awards, the applicant names a sponsor and submits a proposal centered on a research and training plan. Applicants are required to commit at least 75% of their time to K23-supported activities. In 2002, the NCRR will also initiate institutional awards for training in clinical research. These awards, known as K12 awards, will require the recipient to commit at least 90% of their time over a five-year period to training activities. The level of salary support is somewhat higher than that for a K23 award, and, importantly, there is additional support available for didactic training. Many of the recipients of these awards will register for higher degrees, so that they will leave the program with, for example, a Ph.D. Both K23- and K12-supported training programs are likely to encompass laboratory as well as bedside activities, and the future careers of the recipients will be tracked at the NCRR to assess the relative merits of the two training pathways. Both are fully described at http://grants2.nih.gov/training/careerdevelopmentawards.htm.

Research careers are generally less remunerative than, for example, certain subspecialty practices. To reduce this disadvantage, the NIH introduced a Loan Repayment Plan in 2002 that used a contract mechanism to pay off as much as $70,000 of medical student loans to qualified applicants. In the first year of operation, National Institutes of Health support was required to qualify for loan repayment, and in subsequent years other sources of research support will satisfy this eligibility requirement. Additional information is available at www.lrp.nih.gov.

POTENTIAL FOR THE FUTURE: STEM CELL RESOURCES

Lymph drains through vessels, and the vessels themselves comprise a delicate network of cells and valves.[2] But for recent research, their organization would seem far too complex to model. In a major recent advance, Levenberg *et al.* have shown that H9 human embryonic stem cells could be driven to express endothelial cell genes, including Pecam and Cadherin, in tissue culture.[3] Embryonic stem cells may one day provide a means to developing lymphatics *in vivo*, and here, too, the NCRR contributes through funding the expansion and distribution of approved human embryonic stem cell lines.

REFERENCES

1. SABIN, F. 1913. The Origin and Development of the Lymphatic System. Johns Hopkins Press. Baltimore, MD.
2. CASTENHOLTZ, A. 1991. Structure of initial and collecting lymphatic vessels. *In* Lymph Stasis: pathophysiology, diagnosis and treatment. W.L. Olszewski, Ed. CRC Press. Boca Raton, FL.
3. LEVENBERG. S. *et al.* 2002. Endothelial cells derived from human embryonic stem cells. Proc. Natl. Acad. Sci. USA **99:** 4391–4396.

An Overview of the Pathology and Approaches to Tissue Engineering

ERIN R. OCHOA[a] AND JOSEPH P. VACANTI[b]

[a]*Department of Pathology, Montefiore Medical Center, Albert Einstein College of Medicine, Bronx, New York, USA*

[b]*Department of Surgery, Massachusetts General Hospital, Harvard Medical School, Boston, Massachusetts, USA*

ABSTRACT: In tissue engineering, there is an attempt to culture living tissues for surgical transplantation. *In vitro* and *in vivo* approaches have produced vascular and cardiovascular components, cartilage, bone, intestine, and liver. Attempts to microdesign cell-culture support scaffolds have used a new generation of biocompatible and bioabsorbable polymers. Suspensions of donor cells are seeded onto protein-coated polymer scaffolds and grown to confluence in dynamic bioreactors. *In vitro* techniques produce monolayers of tissues. Denser masses are achieved by implanting monolayers onto a host, or by culturing cell/polymer constructs *in vivo*. Existing techniques have produced functioning heart valves from sheep endothelial cells and myofibroblasts. Cultured ovine arterial cells have replaced 2-cm segments of pulmonary artery in lambs. Chondrocyte cultures have produced a human-ear-shaped construct, temporo-mandibular joint discs, meniscal replacement devices, and human-phalange-shaped constructs, complete with a joint. The culture of composite tissue types has recently been reported. Intestinal organoid units containing a mesenchymal core with surrounding polarized epithelia have been used in lieu of an ileal pouch in Lewis rats, and the long-term culture of rat hepatocytes has revealed cellular differentiation and neomorphology resembling elements of a biliary drainage system. To sustain the *in vitro* culture of dense tissues prior to implantation, micro-electro-mechanical systems (MEMS) fabrication technologies have been adapted to create wafers of polymer containing sealed, branching, vascular-type spaces. After seeding with rat lung endothelial cells, followed by 5 days of bioreactor culture, the result is an endothelial network with controlled blood flow rates, pressure, and hematocrit. When these customized vascular systems can be used to support *in vitro* culture, a new generation of dense, composite, morphologically complex tissues will be available for clinical development.

KEYWORDS: tissue engineering; bone cartilage; cardiovascular; MEMS; gastrointestinal; liver; microfabrication

---

Address for correspondence: Joseph P. Vacanti, M.D., John Homans Professor of Surgery, Department of Surgery, Massachusetts General Hospital, Warren 11, 14 Fruit Street, Boston, MA 02114. Voice: 617-724-1725; fax: 617-726-5057.

jvacanti@partners.org

Ann. N.Y. Acad. Sci. 979: 10–26 (2002). © 2002 New York Academy of Sciences.

The emerging field of tissue engineering is best defined by a general statement of its clinical goals: the rational design and fabrication of living human replacement devices for surgical transplantation and reconstruction. In its approximately 20 years of development, tissue engineering has generated a range of strategies for fabricating many human tissue types. Most tissue engineering strategies involve combining a small amount of donor tissue with a support structure. Then, either new transplantable tissue is cultured *in vitro* or the device is implanted into a host, where new tissue can form with neovasculature and signaling factors marshaled from the host.[1] In a relatively brief time, tissue engineering has undergone explosive growth and is now yielding clinical results. Versatile skin replacement devices have been commercially available for some time.[2,3] The fabrication of bone and cartilage devices is well advanced, while vascular and cardiovascular devices are entering the human trial stage.[4–7] This generation of successful devices has been facilitated by the perfection of a range of techniques for fabricating thin layers of tissues with low oxygen and nutrition requirements. The mature phase of this generation of devices can be characterized by the customization of existing techniques to specific clinical scenarios. The next generation of tissue engineering devices will depend upon new techniques for culturing dense and complex tissue masses. Recent advances in the fabrication of intestine and liver, in particular, give us some indication of how the field will develop. The next milestone is likely to be reached when new vascular systems can be employed to sustain the long-term *in vitro* culture of dense, complex tissues prior to implantation.[8]

In meeting the demands of a specific clinical scenario, the current generation of tissue engineering projects begins with a number of strategic decisions regarding cell sourcing, support materials, and the regulation of the mechanochemical factors which guide tissue development.[9–11] Although much attention has been given to the utility of stem cells, a variety of autologous and allogenic sources have supplied successful devices. As we shall see, a growing body of experimental evidence indicates that the answer to most cell sourcing issues may be found in identifying and isolating the smallest functional units of a desired adult tissue type. In other words, it appears that tissue engineering might ultimately be able to avoid using problematic fetal tissue or stem cell technologies.

Although naturally occurring collagen, in the forms of foams, gels, and meshes, has been used to sustain cell culture, new biocompatible and biodegradable polymer scaffolds coated with cell binding and growth-promoting proteins offer greater consistency and versatility in designing microstructures and tightly controlling such properties as mechanical strength and degradation time.[12,13] Polymer scaffolds are particularly useful in culturing tissues for which gross morphology plays the central functional role, such as with bone, cartilage, and ductal structures. Finally, a variety of dynamic bioreactor systems have been developed that simulate a number of physiological conditions such as oxygen exchange, nutrient flow dynamics, and electrical and mechanical stimuli.[14] Alternatively, animal hosts have been used to support growing devices that are implanted into such vascularized spaces as the omentum, mesentery, interscapular fat pad, or latissimus dorsi.[15] Up to the present, the most advanced successes have utilized the latter approach. This indicates that the central technical challenge to *in vitro* culture is to supply greater amounts of oxygen and nutrients to growing tissues. When *in vitro* technology can match *in vivo* culture in this respect, tissue cultures will be indefinitely sustainable, and a host of problems that arise from using a living host as a tissue factory can be avoided.

BONE AND CARTILAGE

Various bone and cartilage devices have been fabricated from chondrocytes isolated from articular surfaces or periosteum. Chondrocytes harvested from articular surfaces tend to differentiate in culture to cartilage, while those isolated from periosteum initially resemble cartilage and then progress to bone.[16] When chondrocytes are seeded onto pre-shaped polymer scaffolds, the results are often dramatic, resulting in physiologically accurate human ear, temporo-mandibular joint disc, and meniscal-shaped constructs grown within host animals.[17–19] More complex devices include cartilaginous tubes lined with respiratory epithelium for use as a tracheal replacement, and the bony repair of cranial and femoral shaft defects in nude rats.[20,21] The most complex device, containing bone and cartilage that resembled a human distal and middle phalanx with an interphalangeal joint, was cultured for 20 weeks subcutaneously in athymic mice.[22] These successes indicate that, in general, the principal functional role of bone and cartilage is mechanical/structural, and that the functional structural unit, the adult chondrocyte, is all that is required to reproduce this range of structures. The main challenges for bone and cartilage tissue engineering are to increase the strength and density of engineered tissues, and to explore specific clinical applications of this mature technology.

CARDIOVASCULAR TISSUE ENGINEERING

Cardiovascular tissue engineering has made progress towards producing functioning heart valves and conduits. Sheep endothelial cells and myofibroblasts seeded onto polymer scaffolds and cultured *in vitro* in a bioreactor configuration that approximated systolic pulse pressure, functioned as a sheep heart valve for 5 months *in vivo*.[23] In the most successful cardiovascular application of tissue engineering, an occluded human pediatric pulmonary ar-

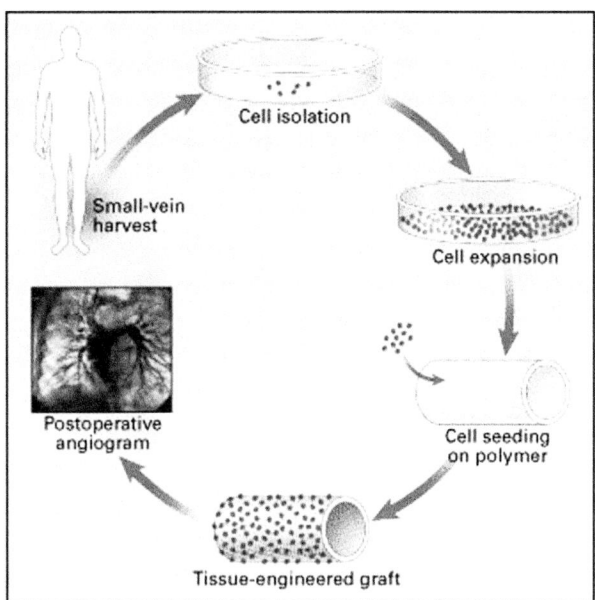

FIGURE 1. Tissue-engineering technique. Venous-wall cells were isolated and expanded *in vitro* and seeded on a biodegradable polymer scaffold. The construct of cells and polymer was implanted as autologous tissue. (From Shinoka *et al.*[24] Reproduced by permission from the *New England Journal of Medicine.*)

tery was reconstructed with a tissue-engineered device produced by culturing, for 10 days prior to implantation, autologous tissue from a peripheral vein on a 10-mm diameter, 20-mm length, 1-mm thick polymer conduit, designed to biodegrade within 8 weeks. After 7+ months, the patient was doing well (FIG. 1).[24] As with bone and cartilage, many vascular and cardiovascular components can be reproduced by concentrating on the gross morphology of the desired tissue. Adult structural units seeded onto polymers that mimic the desired shape can reproduce viable, transplantable tissue components, given adequate nutrition and oxygen.[24] Nevertheless, cardiovascular tissue engineering must improve the strength and viability of constructs as it explores specific clinical scenarios.

GASTROINTESTINAL TISSUE ENGINEERING

Success has also been made in engineering devices whose physiological functions go beyond their mechanical/structural properties. In particular, large and small intestine have been produced by culturing organoid units of Lewis rats.[25] An intestinal organoid unit is a multicellular aggregate contain-

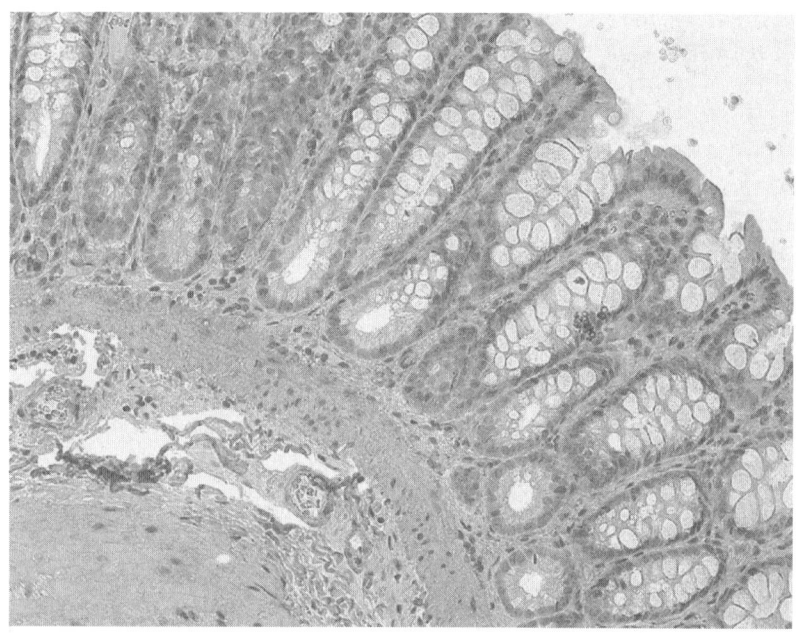
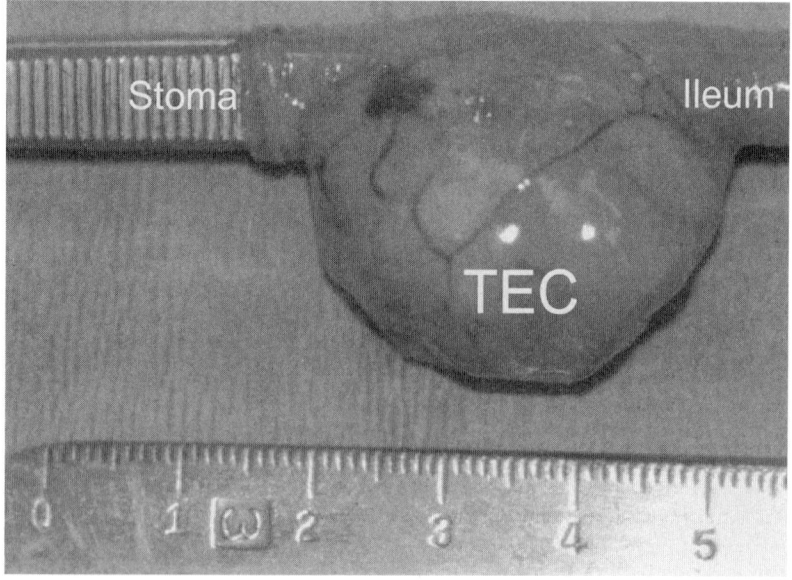

FIGURE 2. Tissue-engineered colon. Gross photograph and 20× magnification. (From Grikscheit et al.[26] Reproduced by permission from *Surgery*.)

ing a mesenchymal core surrounded by polarized intestinal epithelium and containing all of the cells of a full-thickness intestinal section. These are obtained by dissecting the sigmoid colon of a Lewis rat. After purification by enzymatic digestion, titration, and centrifugation, the donor tissue is seeded onto 2-mm polymer conduits and immediately implanted into the omentum of a host rat. Such engineered constructs grow to resemble native colon, with epithelial cells facing inward towards the lumen of the cylinder, reconstitution of other layers of intestinal wall, and substantial vascularization (FIG.2).[26] It is most significant that similar successes have been reported for the culture of organoids obtained from neonatal, adult, and also tissue-engineered sources. Tissue-engineered colon has been anastomosed onto adult male Lewis rat small intestine after 75% bowel resection. The hosts then displayed weight gain, bowel patency, and statistically significant increases in engineered intestine size, with noted gross fluid absorption.[27] In this case, gross morphology, the ductal shape of the desired tissue, is only one of the central functional properties of the desired tissue. The familiar polymer tubes are adequate to reproduce this feature. The other central functional feature is the absorptive properties of the multilayered tissue. These properties are reproducible by culturing complex organoids, rather than a single structural element, such as a chondrocyte, in the case of bone and cartilage, or endothelial cells in vascular conduits.

TISSUE ENGINEERING OF THE LIVER

Although progress has been made in culturing hepatic masses retaining such functions as albumin and transferrin production, the successful tissue engineering of viable, transplantable liver is likely to follow the development of vascular systems that can indefinitely sustain the *in vitro* culture of dense tissue types with high oxygen and nutrition requirements (FIG. 3).[28,29] Therefore, prior to detailed consideration of liver applications, our Microvascular Project should be discussed.

MICROFABRICATION

At the heart of our core tissue-engineering platform is the adaptation of a process developed by the integrated circuit industry for manufacturing microelectro-mechanical systems (MEMS). Working with Draper Laboratory, which has extensive experience in MEMS fabrication, we are fine-tuning a technique that entails first masking a silicon or Pyrex wafer with a micro-patterned, etch-resistant template using photolithography, then etching the exposed pattern of silicon or Pyrex with an anisotropic dry plasma system. After

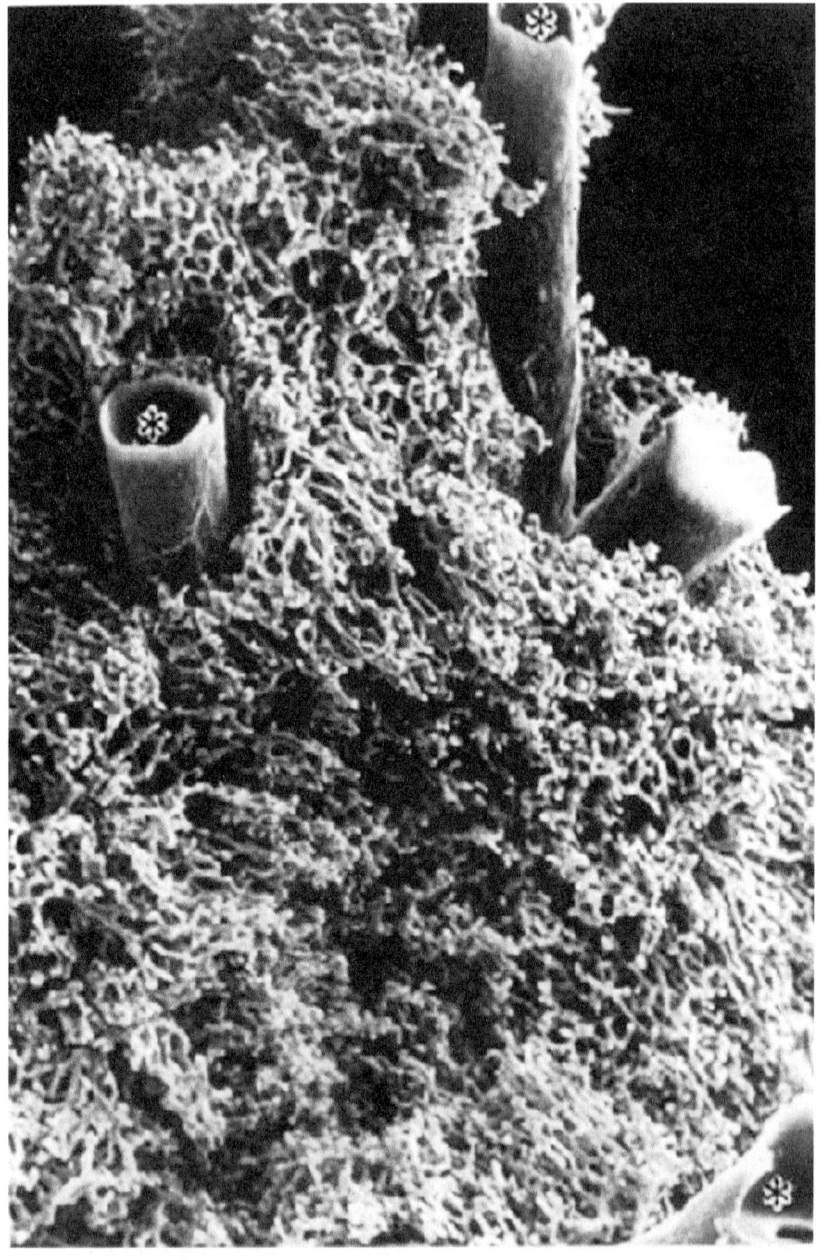

FIGURE 3. Vasculature of the human liver. (From Vonnahme.[29] Reproduced by permission from S. Karger.)

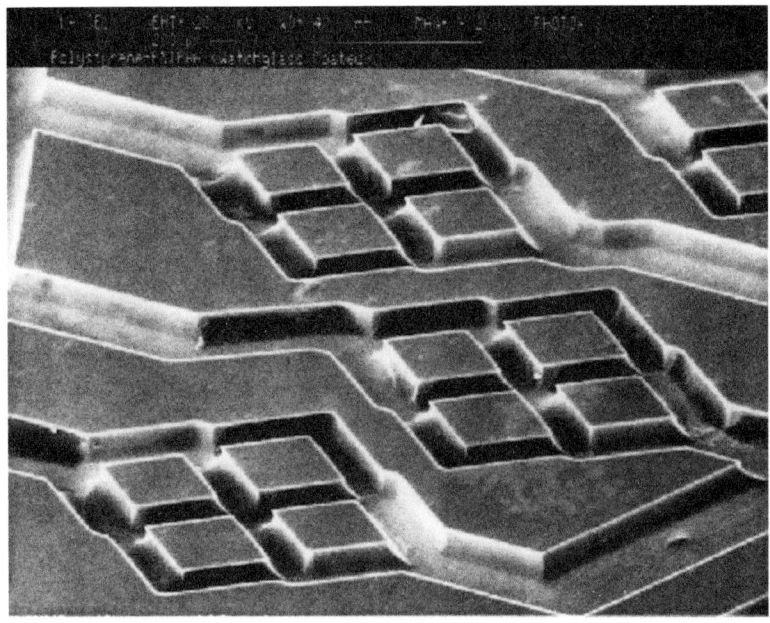

FIGURE 4. Micro-etched wafer. (Courtesy of Dr. J. Borenstein.)

chemically removing all pattern-transfer materials, the resulting micro-etched wafer is then used as a master mold for pattern transfer to a biocompatible polymer (FIG. 4).[30]

Prior to working with this MEMS technology, we had employed a succession of polymer devices, beginning with a non-woven mesh of polyglycolic acid fibers, 15 μm in diameter, and proceeding through several polymer sponges of varying textures and porosities. All of these devices possessed uneven culture-flow characteristics that could only be roughly controlled.[31] With our new MEMS technology, we have produced a series of biocompatible polymer scaffolds whose smooth, branching, two-dimensional architecture is designed to mimic the fluid dynamic properties of the specific vascular structure of any given tissue type (FIG. 5).[32]

Our early tissue-engineering strategies employed a variety of techniques for seeding donor cells and culturing new tissue on our various polymer scaffold prototypes.[14] We have gradually developed a bioreactor system to provide much more effective dynamic cell seeding, to provide continuous culture flow conditions for expanding new tissue, and to conduct fluid dynamic studies of our new two-dimensional sealed vascularized systems (FIG. 6). In our current bioreactor configuration, a medium of isolated vascular endothelial cells is pumped through the sealed, protein-coated polymer network with a low pulsate flow rate and culture viscosity, coordinated to optimize the adhe-

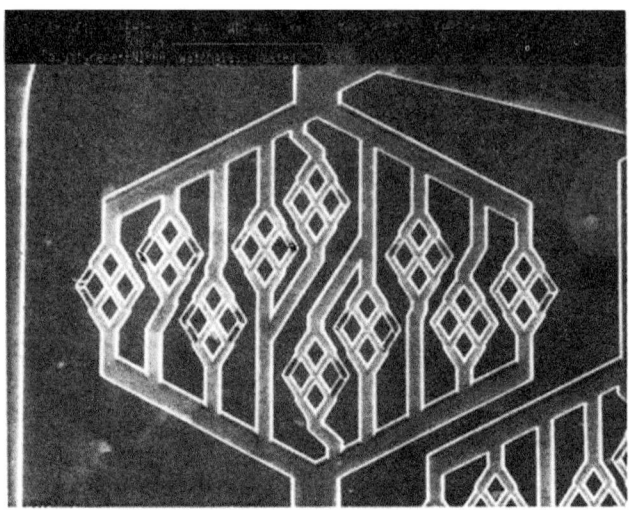

FIGURE 5. Tissue-engineering prototype. (From Kaihara et al.[28] Reproduced by permission from *Tissue Engineering*.)

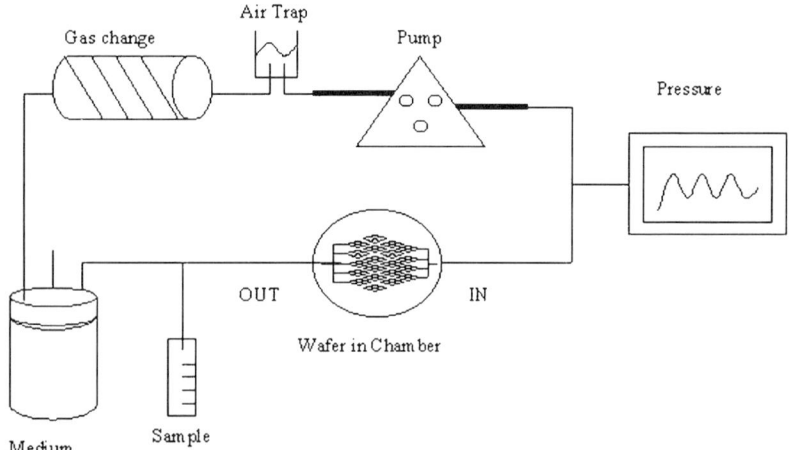

FIGURE 6. Schematic of microcirculation bioreactor system. (Courtesy of Dr. H. Terai.)

sion of cells to the walls of the scaffold without permitting the channels to become occluded. By switching to an oxygen- and nutrient-rich culture medium, we can then maintain flow conditions while seeded donor cells develop into a tubular layer of new tissue lining the scaffold, in about 5 days. The result is a network of living endothelial vessels which mirror our architectural and flow designs in two dimensions, and which demonstrate the ability to transport red blood cells (RBCs), but which are still embedded in polymer.

Once we perfect our etching techniques, we will transition to the fabrication of scaffolds with three-dimensional architecture by bonding together multiple layers of molded polymer. The remaining challenge will be to create strong, sealed three-dimensional bioabsorbable polymer scaffolds coated with an optimal combination of cell-binding and growth factors, and which degrades at a controlled rate as new tissue forms, leaving an intact, living vascular system.

This core vascular platform is being integrated into our long-term Human Liver Project. Three cell types are of particular interest to the tissue engineering of human liver, due to their roles in liver architecture: vascular endothelial cells, hepatocytes, and specialized biliary epithelial cells. Each 2-mm diameter, hexagonal lobule has a dual blood supply. The actual functional subunits or metabolic subunits are referred to as acini. Terminal branches of the hepatic artery and portal vein attach to portal areas of each lobule enclosed by connective tissue (FIG. 7).[33] A one-to-three ratio of oxygen-rich to nutrient-rich blood cells, representing 25% of total cardiac output, mix within these lobules within innumerable, fenestrated sinusoids. RBCs exchange metabolites with surrounding parenchymal cells through the peri-sinusoidal space of Disse, then exit as depleted RBCs through the terminal branch of the hepatic vein, located at the center of each lobule. Together, the terminal portal veins, hepatic arteries, sinusoids, and hepatic veins form one vast, continuous vascular space composed of endothelial cells. Our synthetic networks lined with endothelial cells are our engineering analogue for this complex system.

Surrounding the sinusoids, and extending microvilli into the peri-sinusoidal space of Disse, is a large mass of parenchymal tissue organized into cribiform, anastomosing sheets, whose main functional cell type is the hepatocyte. These sheets radiate from the terminal hepatic artery, defining six roughly triangular parenchymal masses, the acini. Within the acini, three zones are distinguished, graded by the depletion of oxygen and other metabolites in adjacent RBCs as they travel the length of the sinusoid towards the central terminal hepatic vein. Growing a functional mass of liver parenchymal tissue from donor hepatocytes is the primary goal of our Human Liver Project, and the end to which we need our endothelial networks.

Finally, there is another ductal or vascular-type space, completely surrounded by parenchymal tissue, through which hepatocytes continuously secrete bile. A minute, hexagonal bile canaliculus is formed from specialization of the adjacent surfaces of individual hepatocytes, which are sealed with tight junctions. These canaliculi are confluent with terminal ductules that are initially made of squamous cells, but give way to low cuboidal biliary epithelium as they approach the interlobular bile ducts. These ducts are themselves composed of cuboidal to low-columnar epithelium, and are associated with the terminal branches of the hepatic artery and portal vein at three apices of each lobule. One liter of bile is secreted each day by hepatocytes and moved out of the liver by mechanical forces in a direction opposite to the flow of

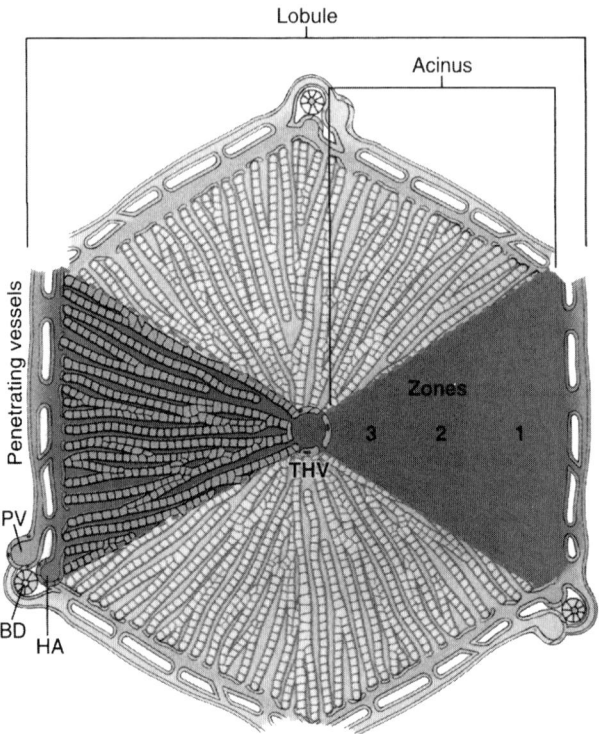

FIGURE 7. Microscopic liver architecture depicted schematically. The classic hexagonal lobule is centered around a central vein (terminal hepatic venule), with portal tracts at three of its apices. The triangular acinus has as its base the vessels that extend from portal veins and hepatic arteries to penetrate the parenchyma. The apex is formed by the terminal hepatic vein. Zones 1, 2, and 3 represent metabolic regions increasingly distant from the blood supply. (From Crawford.[33] Reproduced by permission from W.B. Saunders.)

blood. Engineering this space, composed of specialized biliary epithelial cells, is the secondary goal of our Human Liver Project.

Although insufficient on their own, the techniques that our team has painstakingly acquired, over 15 years, for isolating, seeding, culturing, and transplanting hepatocytes, will guide our initial steps towards the horizon that is emerging as we perfect our engineered vasculature.[13,34] The bulk of our work has been performed with syngeneic donor hepatocytes isolated from adult male Lewis rat livers in a modified two-step collagenase procedure followed by repeated centrifugation.[35,36] A variety of resulting purified single cell suspensions, enriched with hepatocyte growth medium, have been seeded onto a variety of collagen, polymer, silicon, and Pyrex materials (both uncoated and coated with binding proteins) in a variety of architectural configurations, using multiple static and dynamic cell-seeding processes. Our optimal method of dynamic cell seeding takes approximately 90 min to line the support surface uniformly. By repeatedly culturing various cell/matrix combinations in a bioreactor for a range of time periods prior to harvesting and analyzing, we have determined that our optimal suspensions attach to a variety of materials, proliferate, and grow to confluence within 4 days of seeding. Culture has been continued for periods of up to 21 days. The number of live hepatocytes invariable decreases over time; those that survive lose their liver-specific function, and all dense masses contain necrotic cores.[37]

A variety of tissue-engineered hepatocyte structures have also been transplanted onto the omentum of host rats. Our optimal method involves injecting Retrosine into the peritoneal cavity of the host at a dose of 3 mg/100 g at 5 weeks, and again at 3 weeks, prior to implantation. This inhibits the regeneration of normal liver by producing a block in the hepatocyte cell cycle, with an accumulation of cells in late S and/or G_2 phase. Optimal implantation entails creating a portacaval shunt, implanting a hepatocyte mass that has been cultured for 4 to 14 days onto the microvasculature of the rat omentum, and performing a 60% hepatectomy on the host. Such implants have been grown for periods of up to 3 months prior to harvesting and testing.[10]

The engineering challenge that we now face is to combine the long-term culture of hepatocytes with our vascular endothelial systems. Independently, we can create small masses or sheets of hepatocytes that can survive removal from the support scaffold, with folding and manipulation into various configurations, and we can create two-dimensional vascular endothelial networks embedded in biocompatible polymer. When we can remove an intact endothelial vasculature from the polymer, we may be able to layer vascular networks and hepatocyte sheets and continue culture, with oxygen and nutrition to be provided through the newly engineered vasculature. Alternatively, we may be able to seed donor hepatocytes directly onto a living or synthetic vascular system, growing dense new parenchymal tissue in one continuous process on prefabricated vascular systems.

These possibilities leave open the question of how we can meet our secondary goal of engineering a bile duct system lined with specialized biliary epithelial cells. The histology of some of our more successful hepatocyte cultures offers a suggestion. Many specimens contain multiple structures resembling bile ducts after 2 weeks to 1 month of culture. Some stain positive for gamma glutamyl transpeptidase (GGT), an enzyme expressed at high levels in normal hepatic biliary epithelial cells, but not typically detected in hepatocytes. This suggests that the cells composing these structures are similar to differentiated biliary epithelial cells. Transmission electron microscopy also revealed some adjacent cell-cell borders of hepatocyte masses that formed structures resembling hexagonal bile canaliculi, but it is uncertain whether these were confluent with the duct-like structures (FIG. 8).[38]

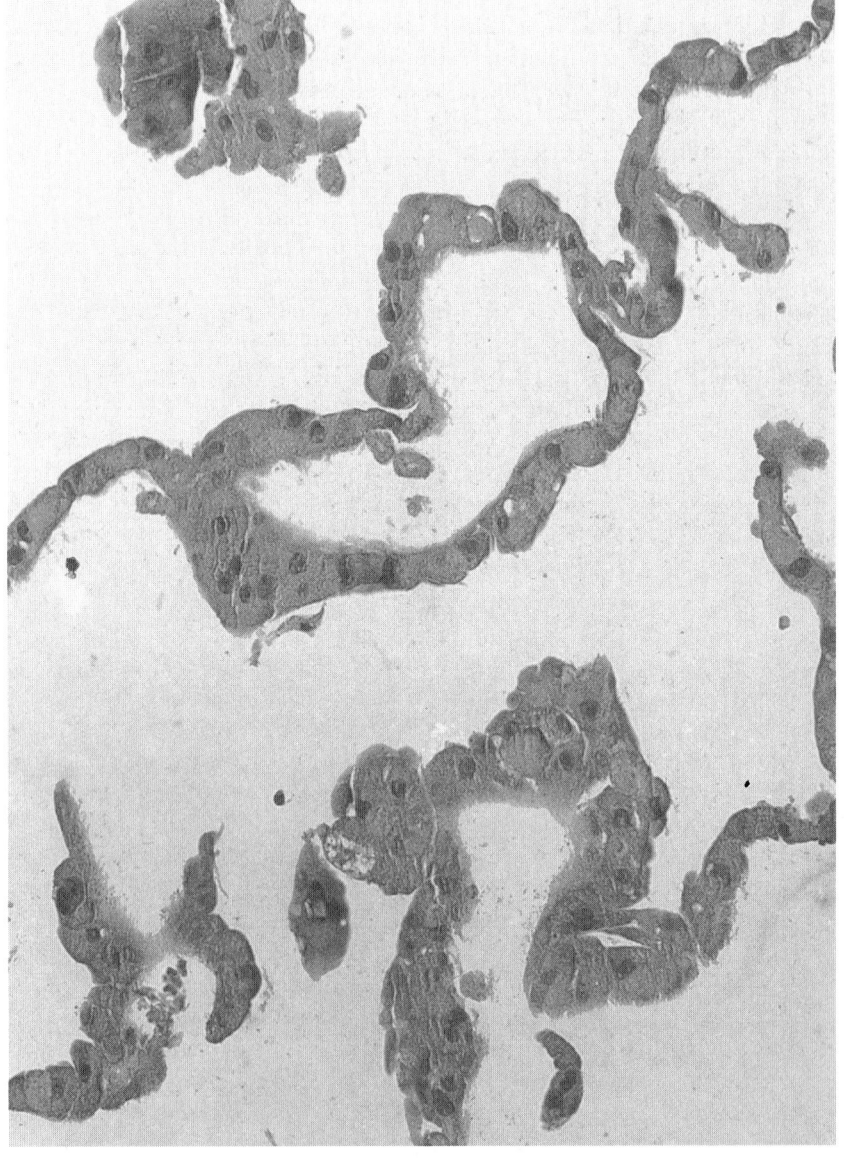

FIGURE 8. Detached monolayers of hepatocytes. Hematoxylin and eosin stain; original magnification ×20. (From Kaihara et al.[28] Reproduced by permission from *Tissue Engineering*.)

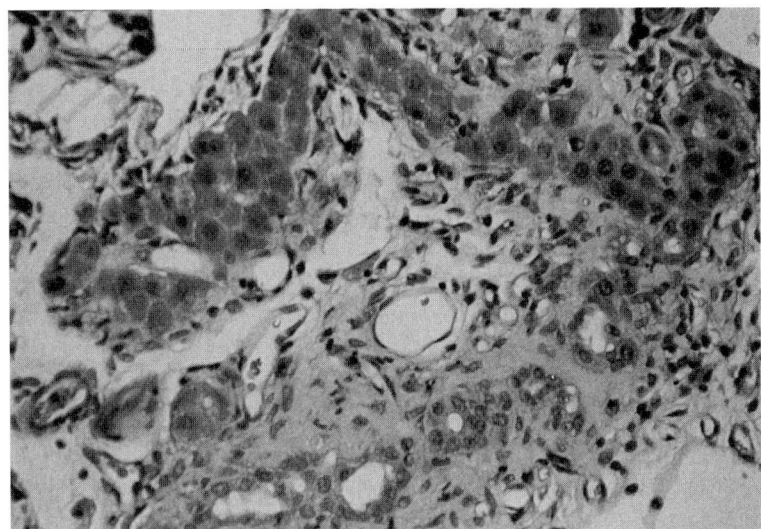

FIGURE 9. Hepatocyte sheets implanted into the rat omentum after 4 weeks. Hematoxylin and eosin stain; original magnification ×20. (Courtesy of Dr. K. Ogawa).

In one case, histology at 2 weeks showed a duct-like structure composed of both cells resembling biliary epithelial and cells that were morphologically more similar to hepatocytes. This bile-duct-like structure was located between the canaliculi-like structures composed of hepatocytes, and the bile-duct-like structures formed solely by cells resembling biliary epithelial, as if it were a transitional structure between the two. This phenomenon suggests the possibility that canaliculi-like structures and bile-duct-like structures may eventually grow to confluence in tissue-engineered constructs (FIG. 9).[39]

These results produce the tantalizing suggestion that a biliary system may spontaneously differentiate and regenerate from our hepatocyte culture once we can sustain the process for longer periods. At this juncture, we return to the presuppositions of our mechanical-structural approach to tissue engineering, namely, that our challenge is to engineer gross structural cues and culture environments, and that, at some point, Nature will take over and complete the process of regenerating whole, functioning liver and other tissues. The near future will tell whether we have attained this goal.

REFERENCES

1. VACANTI, J.P. & R. LANGER. 1999. Tissue engineering: the design and fabrication of living replacement devices for surgical reconstruction and transplantation. Lancet **354:** 32.

2. BELL, E., H.P. EHRLICH, D.J. BUTTLE, *et al.* 1981. Living tissue formed in vitro and accepted as skin-equivalent tissue of full thickness. Science **221**(4486): 1052–1054.
3. NAUGHTON, G.K. 2000. Dermal equivalents. *In* Principles of Tissue Engineering, 2nd ed. R.P. Lanza, R. Langer & J.P. Vacanti, Eds.: 891–901. Academic Press. San Diego, CA.
4. VACANTI, C.A. & J. UPTON. 1994. Tissue engineered morphogenesis of cartilage and bone by means of cell transplantation using synthetic biodegradable polymer matrices. Clin. Plast. Surg. **21**: 445–462.
5. SCHOEN, F.S. & R.J. LEON. 1999. Tissue heart valves: current challenges and future research perspectives. J. Biomed. Mater. Res. **47**: 439–465.
6. SHINOKA, T., D. SHUM-TIM, P.X. MA, *et al.* 1998. Creation of viable pulmonary artery autographs through tissue engineering. J. Thorac. Cardiovasc. Surg. **115**: 536–545.
7. SHUM-TIM, D., U. STOCK, J. IBRKACH, *et al.* 1999. Tissue engineering of autologous aorta using a new biodegradable polymer. Ann. Thorac. Surg. **68**: 2298–2304.
8. OCHOA, E.R. & J.P. VACANTI. 2001. Developing a core platform for the tissue engineering of vital organs. Transplant. Rev. **15**(4): 184–199.
9. BELL, E. 2000. *In* Principles of Tissue Engineering, 2nd ed. R.P. Lanza, R. Langer & J.P. Vacanti, Eds.: xxv. Academic Press. San Diego, CA.
10. LANGER, R. & J.P. VACANTI. 1999. Tissue engineering: the challenges ahead. Sci. Am. **280**: 62.
11. INGER, D.E. 2000. Mechanical and chemical determinants of tissue development. *In* Principles of Tissue Engineering, 2nd ed. R.P. Lanza, R. Langer & J.P.Vacanti, Eds.: 104. Academic Press. San Diego, CA.
12. MOONEY, D.J. & R.S. LANGER. 1995. Engineering biomaterials for tissue engineering: The 10–100 micron size scale. *In* The Biomedical Engineering Handbook. J.D. Bronzino, Ed. CRC Press. Boca Raton, FL.
13. VACANTI, J.P., M.A. MORSE, W.M. SALTZMAN, *et al.* 1988. Selective cell transplantation using bioabsorbable artificial polymers as matrices. J. Pediatr. Surg. **23**: 3–9.
14. FREED, L.E. & G. VURJAH-NOVAKOVIC. 2000. Tissue engineering bioreactors. *In* Principles of Tissue Engineering, 2nd ed. R.P. Lanza, R. Langer & J.P. Vacanti, Eds.: 147. Academic Press. San Diego, CA.
15. YANNAS, I.V. 2000. In vivo synthesis of organs and tissues. *In* Principles of Tissue Engineering, 2nd ed. R.P. Lanza, R. Langer & J.P. Vacanti, Eds.: 163. Academic Press. San Diego, CA.
16. IBARRA, C., R. LANGER & J.P. VACANTI. 1996. Tissue engineering: cartilage, bone and muscle. *In* Yearbook of Cell and Tissue Transplantation. R. Langa & W. Chick, Eds.: 235–245. Kluwer Academic. The Netherlands.
17. VACANTI, C.A., L.G. CIMA, D. RATKOWSKI, *et al.* 1992. Tissue-engineered growth of new cartilage in the shape of a human ear using synthetic polymers seeded with chondrocytes. Mat. Res. Sec. Symp. Proc. **252**: 367–373.
18. PUELACKER, W.C., J. WISSER, C.A. VACANTI, *et al.* 1994. Temperomandibular joint disc replacement made by tissue-engineered growth of cartilage. J. Oral Maxillofac. Surg. **52**: 1172–1178.
19. IBARRA, C., C. JANETTA, C.A. VACANTI, *et al.* 1997. Tissue-engineered meniscus transplantation. Transplant. Proc. **29**: 986–988.

20. SAKATA, J., C.A.VACANTI, B. SCHLOE, et al. 1994. Traded composite tissue engineered from chondrocytes, tracheal epithelial cells, and synthetic degradable scaffolding. Transplant. Proc. **26:** 3309–3310.
21. LEE, I.W., J.P. VACANTI, J. YOO, et al. 1997. A tissue engineering approach for dural and cranial grafts. Presented at the Congress of Neurological Surgeons. New Orleans, LA.
22. ISOGAI, N., W. LANDIS, T.H. KIM, et al. 1999. Formation of phalanges and small joints by tissue engineering. J. Bone Joint Surg. **81A:** 306–316.
23. STOCK, U.A., M. NAGASHIMA, P.N. KHALIL, et al. 2000. Tissue-engineered valve conduits in the pulmonary circulation. J. Thorac. Cardiovasc. Surg. **119:** 732–740.
24. SHINOKA, T., I. YASUHARU & I. YOSITO. 2001. Transplantation of a tissue-engineered pulmonary artery. N. Engl. J. Med. **344**(7): 532.
25. CHOI, R.S., C. POTHOULAKIS, B.S. KIM, et al. 1998. Studies of brush border enzymes, basement membrane components, and electrophysiology of tissue-engineered neointestine. J. Pediatr. Surg. **33:** 991–997.
26. GRIKSCHEIT, T.C., J.B. OGILVIE, E.R. OCHOA, et al. 2002. Tissue-engineered colon exhibits function in vivo. Surgery **132**(2): 200–204.
27. KAIHARA, S., S.S. KIM, B.S. KIM, et al. 2000. Long-term follow-up of tissue-engineered intestine after anastomosis native small bowel. Transplantation **69:** 1927–1932.
28. KAIHARA, S., J. BORENSTEIN, R. KOKA, et al. 2000. Silicon micromachining to tissue-engineer branded vascular channels for liver fabrication. Tissue Engineering **6**(2): 105.
29. VENNAHME, F.J. 1993. The Human Liver: a Scanning Electron Microscopic Atlas. (From the German Die Leber des Menschen: Rasterelektronenmikroscopischer Atlas.) Karger. Basel.
30. LOVE, J.C., J.R. ANDERSON & G.M. WHITESIDES. 2001. Fabrication of three-dimensional microfluidic systems by soft lithography. MRS Bull. **26:** 523–528.
31. MOONEY, D.J.& J.P. VACANTI. 1993. Tissue engineering using cells and synthetic polymers. Transplant. Rev. **7:** 153.
32. KAAZEMPUR-MOFRAD, M.R., J.P. VACANTI & R.D. KAMM. 2001. Computational modeling of blood flow and rheology in fractal microvascular networks. *In* Computational Fluid and Solid Mechanics. K.J. Bath, Ed.: 864–867. Elsevier Science, Massachusetts Institute of Technology. Cambridge, MA.
33. CRAWFORD, J.M. 1999. The liver and the biliary tract. *In* Robbins Pathologic Basis of Disease, 5th ed. R.S. Cotran, V. Kumar & T. Collins, Eds.: 846. W.B. Saunders. Philadelphia, PA.
34. MOONEY, D.J., S. PARKS, P.M. KAUFMANN, et al.1995. Biodegradable sponges for hepatocyte transplantation. J. Biomed. Mater. Res. **29:** 959.
35. AIKEN. J, L. CIMA, B. SCHLOO, et al. 1990. Studies in rat liver perfusion for optimal harvest of hepatocytes. J. Pediatr. Surg. **25:** 140.
36. SEGLEN, P.O. 1976. Preparation of isolated rat liver cells. Methods Cell Biol. **13:** 29.
37. KIM, S.S., H. UTSUNOMIYA, J.A. KOSKI, et al. 1998. Survival and function of hepatocytes on a novel three-dimensional synthetic biodegradable polymer scaffold with an intrinsic network of channels. Ann. Surg. **228:** 8.

38. KAIHARA, S., K. OGAWA, R. KOKA, et al. 2002. In vitro neomorphogenesis in liver tissue on biodegradable polymer scaffolds under dynamic flow culture. Gastroenterology. In press.
39. OGAWA, K., E.R. OCHOA, K. TANAKA, et al. The generation of functionally differentiated, three-dimensional hepatic tissue from two-dimensional sheets of small hepatocytes and non-parenchymal cells. Work in progress.

Tissue Engineering of the Lymphatic System

LAURA E. NIKLASON,[a] JENNIFER KOH,[b] AND AMY SOLAN[a]

[a]*Departments of Biomedical Engineering and Anesthesia, Duke University, Durham North Carolina 22708, USA*

[b]*Department of Bioengineering, University of California at San Diego, San Diego, California, USA*

> ABSTRACT: The field of tissue engineering has seen tremendous expansion in the last decade. In the last several years, tissue-engineering strategies to treat diseases of skin, cartilage, bone, bladder, blood vessel, tendon, and other tissues have been described. However, tissue-engineering approaches to treat diseases of the lymphatic system are currently nonexistent. We propose that acellular tissues, either native or engineered, could be exploited as a platform for the study of lymphatic biology, and for lymphatic tissue engineering. While speculative, this type of experimental model system could prove powerful for dissecting molecular and cellular events surrounding tumor invasion of lymphatics, as well as lymphangiogenesis. Scaffolds seeded with genetically engineered lymphatic cells could also be implanted to repopulate lymphatic vasculature. In the future, the lymphatic system will surely be added to the list of tissues and organs that prove amenable to tissue-engineering therapies.
>
> KEYWORDS: tissue engineering; lymphatic system mechanics

INTRODUCTION

The field of tissue engineering has seen tremendous expansion in the last decade. Although it has been variously defined, tissue engineering may be envisioned as the combination of mammalian cells with extracellular scaffolding material, to produce tissue-like composites for scientific, diagnostic, or therapeutic purposes. In the last several years, tissue-engineering strategies to treat diseases of skin, cartilage, bone, bladder, blood vessel, tendon, and other tissues have been described. A few of these strategies (treatments

---

Address for correspondence: Laura E. Niklason, M.D., Ph.D., Department of Biomedical Engineering, Duke University, Room 136, Hudson Hall, Research Drive at Science Drive, Durham, North Carolina 27708. Voice: 919-660-5149; fax: 919-684-5577.
nikla001@mc.duke.edu

for skin and cartilage) are FDA-approved or in clinical trials, while the vast majority are still at the preclinical or *in vitro* stage. However, tissue-engineering approaches to treat diseases of the lymphatic system are currently nonexistent.

Cells of the lymphatic system, combined with suitable scaffolds or extracellular matrix components, could provide powerful tools for both scientific inquiry and potentially, in the future, for therapy of lymphatic disorders. We will outline how tissue-engineering approaches may be used to study the molecular events surrounding lymphatic vessel formation *in vitro*, to query the interactions between lymphatics and tumor cells, and perhaps eventually to provide cell-based therapies for diseases such as lymphedema.

DESIGN CRITERIA FOR ENGINEERING LYMPHATICS

Mechanical Considerations

Lymphatic vessels serve to conduct extracellular fluid that has been extravasated from the blood vascular system back into the circulation. Lymphatics are embedded in a connective tissue matrix that provides many of their passive mechanical properties,[1,2] although smooth muscle cells in the lymphatic wall also produce active contractions.[3]

When considering the mechanics of the lymphatic system, it is useful to compare them with the blood vascular system.[4,5] The mechanics of blood vessels have been studied for several decades, and engineered blood vessels have achieved mechanics that approximate those of native arteries.[6,7] Hence, comparison of the mechanics between lymphatics and blood vessels lends perspective to the design requirements for lymphatic vessels.

The tensile properties of lymphatics and blood vessels differ markedly (FIG. 1). The walls of muscular arteries contain substantial quantities of collagen and elastin. These densely packed extracellular matrix proteins confer tensile strength to arteries, and result in Young's moduli (E) on the order of 1–100 MPa.[8] In contrast, lymphatic vessels are composed of thin cell layers that are connected to the surrounding loose interstitial matrix via anchoring filaments (FIG. 1). Lymphatics can tolerate applied pressures on the order of 20 mmHg,[2] which is much less than the 2,000 mmHg that muscular arteries can withstand. Correspondingly, moduli for lymphatics are several orders of magnitude less than those of arteries.

From these observations, it is clear that reaching mechanical and structural goals for engineering lymphatics should be much easier than for engineering blood vessels. Given that blood vessels have already been engineered with mechanics far superior to those of lymphatics, this part of the problem appears tractable.

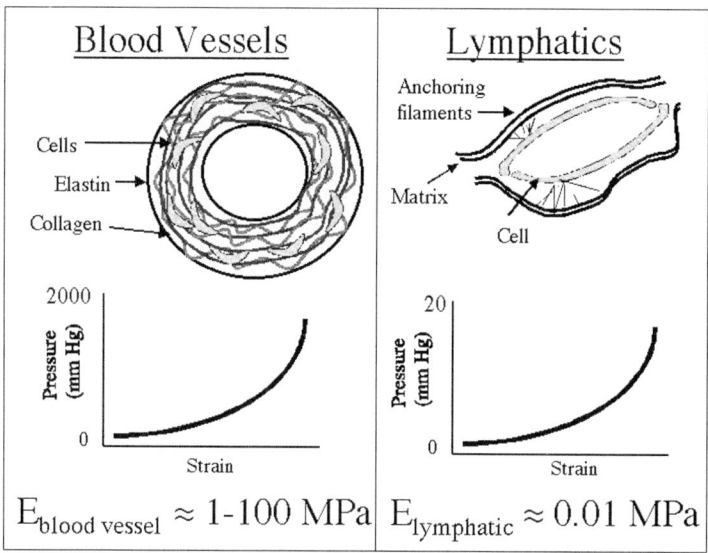

FIGURE 1. Vascular and lymphatic mechanics. *Left:* Muscular blood vessels are composed of vascular cells and a dense elastin/collagen network. Young's moduli (E, slope of stress-strain curve) can vary from 1–100 MPa, depending on the percent strain and the degree of smooth muscle activation. *Right:* Lymphatics, in contrast, are supported by loose interstitial matrix, and tolerate much lower pressures than blood vessels. Young's moduli are on the order of 0.01 MPa.

Endothelial Biology

Lymphatic and vascular endothelial cells are morphologically and functionally quite similar. Until recently, this fact rendered the selective study of lymphatic endothelium quite difficult. Both vascular and lymphatic endothelia express von Willebrand factor, anti-thrombin III, platelet-endothelial adhesion molecule, and CD-34.[9] However, lymphatic endothelium also selectively expresses vascular endothelial growth factor receptor-3 (VEGFR-3), which is the receptor for VEGF-C. VEGF-C has been implicated as a stimulus for lymphangiogenesis.[10] In addition, lymphatic endothelium expresses LYVE-1, a recently discovered receptor for hyaluronic acid that is not present on vascular endothelial cells.

For a tissue-engineered blood vessel to be functional, the engineered endothelium should possess many of the native anti-inflammatory and anticoagulant molecules that are expressed *in vivo* (FIG. 2). Native endothelium synthesizes multiple antithrombotic molecules, which normally inhibit both platelet function and the coagulation cascade.[11,12] Maintenance of an anticoagulant phenotype, including expression of thrombomodulin, nitric oxide,

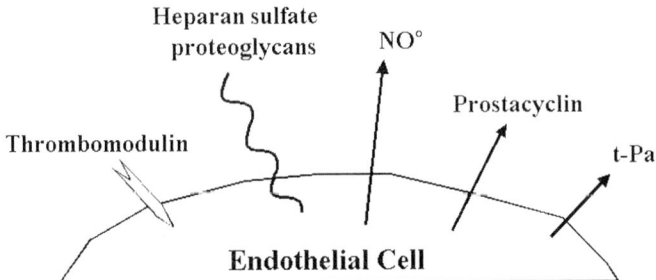

FIGURE 2. Vascular endothelium expresses multiple substances that inhibit coagulation. Thrombomodulin activates protein C; heparan sulfate proteoglycans activate antithrombin III; nitric oxide (NO) and prostacyclin inhibit platelets; and t-Pa initiates clot lysis. Corresponding functions in lymphatic endothelium are less well understood.

prostacyclin, and tissue plasminogen activator (t-Pa), is critical to avoid thrombosis in implanted vascular grafts. When endothelium becomes inflamed or injured, a more pro-coagulant and leukocyte-binding phenotype is expressed, which can impair vascular graft function.

With respect to engineered lymphatic vessels, we anticipate that similar endothelial requirements would apply. Prevention of fibrin plugging by endothelial cells in lymphatics is probably functionally important *in vivo*. In addition, lymphatic endothelium doubtless interacts with lymphocytes via multiple cell-specific pathways. However, our understanding of the leukocyte-interactive and anticoagulant properties of lymphatic endothelium is at an extremely early stage. Hence, while we have a detailed understanding of vascular endothelium, we do not have nearly a comparable understanding of lymphatic endothelium. Advances in our knowledge of molecular function of lymphatic endothelium will greatly facilitate both the study of lymphangiogenesis and lymphatic tissue engineering.

PROGRESS IN TISSUE ENGINEERING

Recently, several groups have reported progress in the development of tissue-engineered blood vessels.[6,7,13,14] We have developed strategies to grow autologous tissue-engineered arteries from vascular smooth muscle and endothelial cells. Vascular cells are seeded onto highly porous, degradable polymer scaffolds. Tubular scaffolds are suspended in bioreactors, which provide pulsatile mechanical forces to developing vessels. During 8 weeks of culture, the polymeric scaffolding substantially degrades, and the vascular cells replicate to form a dense and confluent tissue (FIG. 3A). Smooth muscle cells deposit extensive collagenous matrix, which confers mechanical integrity to the

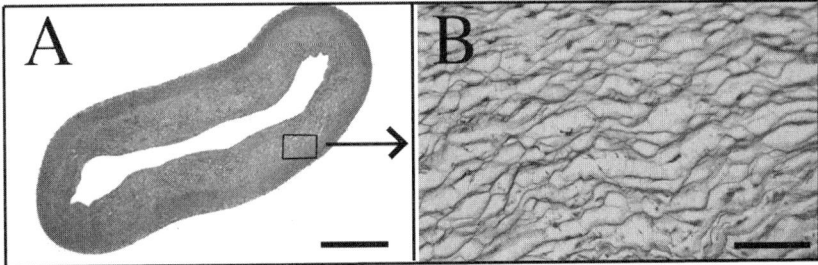

FIGURE 3. Engineered blood vessels. (A) Engineered bovine vessel; hematoxylin and eosin stain. Scale bar = 500 microns. (B) Decellularized engineered bovine vessel; hematoxylin and eosin stain. Scale bar = 50 microns.

engineered constructs. Vessels grown using these methods have rupture strengths that are comparable to native arteries, and are functional when implanted into animals.

Interestingly, engineered blood vessels may be subjected to techniques developed for native tissue decellularization (FIG. 3B). Decellularization strategies are commonly applied to native tissues to produce mechanically strong, acellular implants for surgical reconstruction.[15,16] Acellular implants retain extracellular matrix, but lack cellular components that stimulate immune rejection by the host. Decellularization strategies typically employ combinations of high-ionic-strength salts, detergents, and proteases to selectively remove cellular material. The resulting constructs, at the gross level, appear unchanged from the original specimens, but constitute a structured extracellular matrix onto which other cell types can be seeded and grown.

We propose that acellular tissues, either native or engineered, could be exploited as a platform for the study of lymphatic biology. Interstitial collagenous matrices, such as the decellularized engineered vessel in FIGURE 3B, are ideal substrates for lymphatic cell culture. This is because fibrillar collagen materials are far more biomimetic than standard two-dimensional tissue-culture techniques, or even denatured collagen gels.

Although we have not yet cultured lymphatic cells on decellularized matrices, we have shown the feasibility of repopulating decellularized tissues with vascular cells. A red fluorescent cell linker, PKH26 (Sigma, St. Louis, MO), was utilized to fluorescently label porcine vascular smooth muscle cells in culture. A native porcine carotid artery was decellularized, using 1 M NaCl, 25 mM EDTA, and 1.8 mM sodium dodecylsulfate (SDS). PKH26-labeled smooth muscle cells were incubated with the decellularized interstitium to allow cellular attachment, and then the construct was cultured for 72 hours.

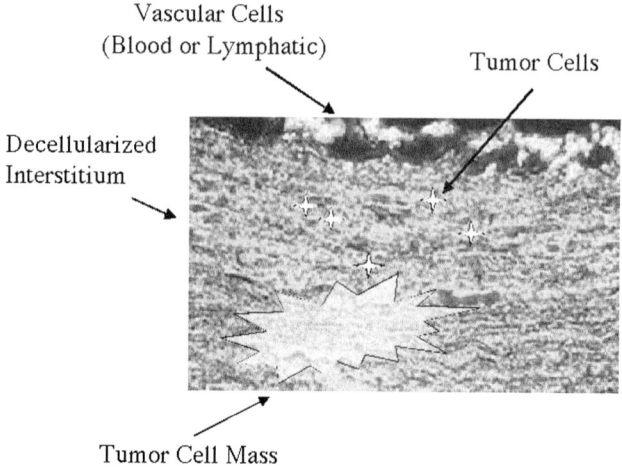

FIGURE 4. Tissue-engineering model system. Fluorescence micrograph of red-staining vascular cells that are seeded onto a decellularized matrix. Scenario of co-culture with tumor cells is illustrated. Note the tumor cell mass (large starburst) in the lower left of the image. Tumor cells (small star shapes) appear towards the center of the image. Vascular cells appear towards the top of the image, while the bulk of the image depicts decellularized interstitium.

A fluorescence micrograph of a decellularized interstitium, partially repopulated with red-staining vascular cells, is shown in FIGURE 4. Also shown diagrammatically in this figure is one possible experimental scenario wherein both tumor and (blood or lymphatic) vascular cells are grown together. While clearly speculative, this type of experimental model system could prove powerful for dissecting molecular and cellular events surrounding tumor invasion of lymphatics, as well as lymphangiogenesis.

FUTURE WORK

Tissue-engineering systems, such as those described above, could permit the biomimetic reconstitution of lymphatic capillaries *in vitro* from lymphatic endothelial cells, or possibly from precursor cells. Events accompanying lymphangiogenesis, such as the interactions of endothelial cells with matrix via anchoring filaments, may be studied in detail if decellularized interstitium is employed. We may inquire as to the effects of culture environment on lymphatic cells. For example, is differentiated phenotype (that is, VEGFR-3 expression) altered in a fibrillar collagen matrix as compared to fibrin or collagen gels? This type of three-dimensional culture system will also allow us to pinpoint soluble factors that are critical for lymphangiogenesis *in vitro*. What is the role of growth factors such as basic fibroblast growth factor, in addition to VEGF-C, in lymphatic formation? Are other agents that are implicated in capillary stability, such as the angiopoetins, important for lymphatic stability?

Tissue-engineering systems also permit mimicry of events surrounding lymphatic spread of tumor, using co-culture of lymphatics and tumor cell populations. Tumor expression of VEGF-C, or other soluble factors, may be confirmed to promote increased local lymphatic density. What other chemotactic or lymphangiogenic factors are secreted by tumor cells, and do these factors modulate the cell adhesiveness of lymphatic endothelium to promote metastasis? Using labeled tumor cells and confocal microscopy techniques, events surrounding tumor cell infiltration into lymphatic vessels can be studied. We may determine if lymphatic metastasis is an active, migratory, and invasive process, or a passive one.

Lastly, in the long term, cellular therapeutic approaches may be developed to treat lymphatic disease. In congenital lymphatic hypoplasia leading to lymphedema, gene therapy to reconstitute functional VEGFR-3 expression in autologous lymphatic endothelium is already being considered. Scaffolds seeded with genetically engineered lymphatic cells could also be implanted to repopulate lymphatic vasculature. It is likely that such approaches would need to be coupled with controlled release of cytokines that are found to be critical for formation and stabilization of lymphatics. In the future, the lymphatic system will surely be added to the list of tissues and organs that prove amenable to tissue-engineering therapies.

REFERENCES

1. LEAK, L.V. & J.F. BURKE. 1966. Fine structure of the lymphatic capillary and the adjoining connective tissue area. Am. J. Anat. **118:** 785–810.
2. OHHASHI, T. 1987. Comparison of viscoelastic properties of walls and functional characteristics of valves in lymphatic and venous vessels. Lymphology **20:** 219–223.
3. ZAWIEJA, D.C., K.L. DAVIS, R. SCHUSTER, *et al.* 1993. Distribution, propagation and coordination of contractile activity in lymphatics. Am. J. Physiol. **264:** H1283–H1291.
4. ARMENTANO, R.L., J.G. BARRA, J. LEVENSON, *et al.* 1995. Arterial wall mechanics in conscious dogs: assessment of viscous, inertial, and elastic moduli to characterize aortic wall behavior. Circ. Res. **76:** 468–478.
5. BARRA, J.G., R.L. ARMENTANO, J. LEVENSON, *et al.* 1993. Assessment of smooth muscle contribution to descending thoracic aortic elastic mechanics in conscious dogs. Circ. Res. **73:** 1040–1050.
6. NIKLASON, L.E., J. GAO, W.M. ABBOTT, *et al.* 1999. Functional arteries grown *in vitro*. Science **284:** 489–493.

7. L'HEUREUX, N., S. PAQUET, R. LABBE, *et al.* 1998. A completely biological tissue-engineered human blood vessel. FASEB J. **12:** 47–56.
8. ARMENTANO R.L., J. LEVENSON, J.G. BARRA, *et al.* 1991. Assessment of elastin and collagen contribution to aortic elasticity in conscious dogs. Am. J. Physiol. **260:** H1870–H1877.
9. SWARTZ, M.A. 2001. The physiology of the lymphatic system. Adv. Drug Delivery Rev. **50:** 3–20.
10. PEPPER, M.S. 2001. Lymphangiogenesis and tumor metastasis: myth or reality? Clin. Cancer Res. **7:** 462–468.
11. WU, K.K. & P. THIAGARAJAN. 1996. Role of endothelium in thrombosis and hemostasis. Annu. Rev. Med. **47:** 315–331.
12. ESMON, C.T. 2000. Regulation of blood coagulation. Biochim. Biophys. Acta **1477:** 349–360.
13. NIKLASON, L.E., W.A. ABBOTT, J. GAO, *et al.* 2001. Morphologic and mechanical characteristics of bovine engineered arteries. J. Vasc. Surg. **33:** 628–638.
14. HUYNH, T., G. ABRAHAM, J. MURRAY, *et al.* 1999. Remodeling of an acellular collagen graft into a physiologically responsive neovessel. Nature Biotechnol. **17**(11): 1083–1086.
15. BADER, A., T. SCHILLING, O.E. TEEBKEN, *et al.* 1998. Tissue engineering of heart valves—human endothelial cell seeding of detergent acellularized porcine valves. Eur. J. Cardio-Thoracic Surg. **14:** 279–284.
16. SCHMIDT, C.E. & J.M. BAIER. 2000. Acellular vascular tissues: natural biomaterials for tissue repair and tissue engineering. Biomaterials **21:** 2215–2231.

Part 1: New Tools for Lymphatic Investigation
Panel Discussion

Lymphatic-Specific Architecture in Tissue-Engineered Vessels

STANLEY G. ROCKSON (*Stanford University*): Is it conceivable that the somewhat loose connection of the endothelium might actually prove to be an advantage as one attempts to construct lymphatic vessels in the laboratory?

There exists a very intimate relationship of lymphatic microvessels to the interstitial matrix and they possess a unique architectural ultrastructure with anchoring reticular filaments. Do these considerations pose any conceptual challenges when one contemplates the construction of such vessels in the laboratory?

LAURA NIKLASON (*Duke University*): Yes and no. I believe that the ability to construct within a collagenous interstitium will probably be half the battle. I suspect that the anchoring filaments are most likely synthesized by the lymphatic endothelium and, if we can provide cells that have sufficiently differentiated function, the cells themselves would be able to provide those specific structural characteristics. These considerations are highly conjectural, but would seem to have a high likelihood for success, in view of current successes in other areas.

Vascular Smooth Muscle Function in Engineered Vessels

DAVID ZAWIEJA (*Texas A & M University*): Do your engineered arterioles or small arteries have normal vascular smooth muscle function? Do they respond to pharmacological agents? Do they have a semblance of myogenic tone?

NIKLASON: Yes, in all cases. The vessels have some element of passive tone, and if, in fact, we excise the mature vessels from the bioreactors, they spontaneously constrict, both radially and longitudinally. They also react to a subset of vasoconstrictors and vasodilators, including bradykinin, serotonin, and prostaglandin F2α. They do not respond to the adrenergic amines, such as epinephrine, and phenylephrine. My theory is that they've lost the receptors, but I'm not certain.

ZAWIEJA: What about responsiveness to endothelially mediated dilatory effects, for instance?

NIKLASON: Depending on the quality of the endothelium, we are able to observe acetylcholine-induced relaxation, but the response varies from vessel to vessel.

ZAWIEJA: In lymphatic investigation, the contractile nature of the lymphatics is of interest, because in the lymphatic circulation, there is no cardiac pump, so the pressure must be generated, in many cases, by the vessels themselves: they act as both conduit and pump. It would be interesting to see whether, in lymphatic engineering, you can develop a contractile pump, as well.

Lacteals in Tissue-Engineered Intestine

ZAWIEJA: Dr. Ochoa, when you showed the artificial intestine that you had developed, you showed two comparison slides, the native intestine and the engineered organ. In the native intestine, there was clear evidence of the presence of lacteals in the central portion of the villi. In some of the villi of the engineered intestine, there were similar-appearing structures. Do you know whether there was indeed lacteal investiture of the lymphatics into the villus?

ERIN OCHOA (*Albert Einstein College of Medicine*): It is difficult to differentiate histologically lymphatic from blood capillaries. However, we have positively identified ganglions in these tissue-engineered constructs. We have positively identified smooth muscle of the muscular mucosa of the lamina propria, and there are blood vessels with red blood cells in them. Therefore, it would be very strange if only the lymphatic system were absent. I would assume that it is there, as well. Morphologically, there is a strong resemblance to the lacteal structures. I would think that the lymphatic structures are there, as well, since we have nerves, muscles, mucosa, and specialized epithelial cells.

ZAWIEJA: Presumably, it would be very important for lipid absorption and normal lipid metabolism in the intestine, as well.

GRANGER (*Louisiana State University*): As a follow-up to that point, if those rats are fed a lipid meal, do you see any chylomicron-filled channels in the intestine?

OCHOA: In answer to your question, I say, indirectly, that the serum studies are entirely normal. The feces show no evidence of fat malabsorption. Thus, both the small and large intestinal construct are functioning normally.

BREN GANNON: (*Flinders University*): One other possible indication of lymphatic development in your intestine might be the presence of Peyer's patches, which might be seen macroscopically.

OCHOA: The Peyer's patches are not overwhelming. On the other hand, we have attempted to produce a tissue-engineered spleen, which I think was successful. We haven't attempted lymph nodes.

Venous and Arterial Endothelial Cells in Tissue Engineering

JÖRG WILTING (*University of Göttingen*): Dr. Niklason, we have recently published a paper in which we demonstrated that, in embryos, a venous endothelial cell can be transformed into an arterial one. This depends on where it integrates into the growing embryo. Have you tried to change a vein into an artery *in vitro*?

NIKLASON: No, we don't take that approach. We are working with individual cells. But clearly it's not an unreasonable approach: clearly, venous structures respond to mechanical stimuli, both at the smooth muscle cell level and at the endothelial level.

Identification of Lymphatic Vessels in Engineered Tissues

SIMON SIMONIAN (*Georgetown University*): An indirect method of identifying lymphatics might be to identify the cellular identity of their contents (i.e., lymphocytes), in contrast to arteries and veins, which contain red blood cells.

OCHOA: From a pathologist's perspective, the distinction is not so readily made. One can have red blood cells in lymphatic channels, and lymphocytes are seen in blood vascular channels.

A Research Program for the Immunophysiology of the Lymphatic System

WALDEMAR OLSZEWSKI (*Polish Academy of Sciences*): I am very interested in the policies of the National Institutes of Health, and their vision of the lymphatic system. I am concerned about the minimal comprehension concerning the lymphatic system as a morphological structure of the immune system. I'm concerned that, in the future, there will not be much work done on the role of the lymphatic system in the immune defense—an area in which there is a lot of work to be done. If, for example, we extirpate all the lymphatics in a human extremity, what ensues is not a necrosis of the tissue, but simply fibrosis, which within years leads to the lack of capillary transport of proteins from the interstitial space. There is no transmigration of cells, but still the tissue survives. However, infections are the plague of such patients, not only bacterial and viral, but also Kaposi's sarcoma and even lymphangiosarcoma. Thus, an immunologically privileged place becomes immunologically incompetent. Without a research focus upon these issues, I do not believe that much progress will occur. I am advocating an expansion of the research program at the NIH to encompass the immunophysiology of the lymphatic system.

HAYWARD (*National Institutes of Health*): I want to respond to the overall question of the NIH vision for lymphoid tissue and lymphatics, because I think now is the moment of real opportunity. Let me emphasize, first of all, that the NIH is a responsive vessel. If you stimulate a lymphatic that has smooth muscle, you hope to see it contract. And I think, equally, that you can elicit a response from the NIH. I'm not trying to suggest for a moment that there should be a National Institute of Lymphatics, but I would say that if you wished higher priority to be given to lymphatics, then it is important that you should all be very vocal. Is this the right moment to do it? I would say surely, since the NIH is about to have new leadership.

Tissue Engineering of Lymph Nodes

OLSZEWSKI: I would like to address some comments to Dr. Ochoa. We would, of course, very much like to see this work extended to the lymph nodes. This is probably a much easier task, because there is a lot of regenerative power in various segments of the lymphoid tissue, including with follicular dendritic cells and endothelial cells.

OCHOA: We could use the same technique that we've used for the other organs. One takes a lymph node, chops it up, isolates the stem cells from it, grows it in culture on a polymer, and then implants it in the omentum and observes whether a lymph node has been created. So, it would just mean applying the same techniques that we've used for the other organs, and I don't see why it wouldn't work.

Research Perspectives in Inherited Lymphatic Disease

ROBERT E. FERRELL

Department of Human Genetics, Graduate School of Public Health, University of Pittsburgh, Pittsburgh, Pennsylvania 15261, USA

ABSTRACT: The hereditary lymphedemas provide an opportunity to identify genes involved in normal and deranged lymphatic development. Genetic analysis of families with Milroy's disease identified mutations in VEGFR3 as a cause of congenital lymphedema, confirming the importance of VEGFC/VEGFR3 signaling in lymphatic development. These observations led to the identification of a mouse model for primary lymphedema, and subsequent analysis of this mouse model, using transgenic and gene transfer techniques, has provided initial clues to the development of a biologically based therapy for primary lymphedema. Of more importance from a public health perspective is the fact that manipulation of this pathway may lead to effective therapies for the more prevalent forms of secondary lymphedema. Identification of FOXC2 as the gene mutated in the lymphedema–distichiasis syndrome has revealed new molecular insight into lymphatic development. Molecular analysis of the FOXC2 pathway may provide clues to developmental pathways shared by the lymphatic system and the other developmental abnormalities associated with this complex syndrome. With improving knowledge of the human genome, genetic analysis of families with lymphedema continues to offer one of the most promising approaches to identifying genes influencing lymphatic development.

KEYWORDS: lymphedema; VEGF-C; VEGFR3; FOXC2; lymphedema–distichiasis; Aagenaes' syndrome; Hennekam's syndrome; Noonan's syndrome

Lymphedema is the abnormal accumulation of protein-rich fluid (lymph) in the interstitial spaces as a result of an anatomical or functional defect in the lymphatic vessels or lymph nodes. The diagnosis of lymphedema excludes the edema that ensues from systemic abnormalities, such as heart, kidney, or liver failure, that far more commonly cause edema of the lower limbs. Prima-

---

Address for correspondence: Robert E. Ferrell, Ph.D., Department of Human Genetics, Graduate School of Public Health, University of Pittsburgh, Pittsburgh, Pennsylvania 15261.
rferrell@mail.hgen.pitt.edu

TABLE 1. Mendelian disorders with primary effects on lymphatics

Autosomal Dominant Disorders	McKusick #	Locus	Gene	Features
Hereditary lymphedema I (Milroy's or Nonne-Milroy's lymphedema)	153100	5q34-q35	VEGFR3	Congenital lymphedema of the legs
Hereditary lymphedema II (Meige's lymphedema)	153200	—	—	Peripubertal onset
Lymphedema–distichiasis	153400	16q24.3	FOXC2	Peripubertal onset
Lymphedema and ptosis	153000	—	(FOXC2)	Shares features with McKusick 153400
Lymphedema and yellow nails	153300	—	(FOXC2)	Shares features with McKusick 153400
Lymphedema, micro-cephaly, and chorioretinopathy	152950	—	—	Microcephaly with normal intelligence, congenital lymphedema, progressive visual impairment
Intestinal lymphangiectasia	152800	—	—	
Noonan syndrome	163950	12q24.1	PTPN11	Lymphdema, pterygium colli, congenital heart defects
Cholestasis–lymphedema syndrome (Aagenaes' syndrome)	214900	15q	—	Recurrent cholestosis, lymphatic hypoplasia
Hennekam's lymphangiectasia–lymphedema	235510	—	—	Protein-losing intestinal lymphangiectasia, abnormal facies, mental retardation
Hereditary lymphedema	247440	—	—	Not well defined

ry, or hereditary, lymphedema may occur as an autosomal dominant condition in which lymphedema is the only clinically apparent abnormality (Milroy's disease) or may occur as one manifestation of a more complex genetic syndrome, with either an autosomal dominant or autosomal recessive pattern of inheritance. The major genetic conditions in which lymphedema is the sole or a consistent phenotypic manifestation are given in TABLE 1. The clinical genetic heterogeneity in lymphedema syndromes suggested by TABLE 1 is being confirmed by the emerging molecular genetic studies of these phenotypes.

HEREDITARY LYMPHEDEMA I (MILROY'S DISEASE)

Milroy presented the first extensive description of primary lymphedema segregating as an autosomal trait in a single family.[1] In the study reported by Milroy, there were 22 affected individuals in five generations. Milroy's clinical description, which resembled that of a smaller family reported by Nonne,[2] has since been referred to as Nonne–Milroy's disease or Milroy's disease. Milroy's disease is present at birth or is diagnosed at an early age, may affect one or both limbs, and progresses slowly from a mild painless swelling to a huge swollen extremity. With the relentless progression of edema, the skin becomes hyperkeratotic and the underlying tissues fibrotic. There is an increasing risk of cellulitis and lymphatitis. Milroy's disease is not associated with congenital anomalies or mental retardation, and subsequent descriptions of numerous families confirmed the autosomal dominant pattern of inheritance with an early age at onset (<3 years) and variable penetrance. Anatomically, Milroy's disease is characterized by hypoplasia or aplasia of the lymphatics. Primary lymphedema has long been recognized as a developmental disorder of the lymphatics, but only recently have the molecular tools become available to characterize the underlying pathogenesis.

Ferrell *et al.* used a genetic approach to localize a gene causing early onset primary lymphedema in three families,[3] and this localization was quickly confirmed.[4,5] Ferrell *et al.* reported a nonconservative, leucine to proline, mutation in the positional candidate geneVEGFR3 (vascular endothelial growth factor receptor-3; Flt4) in one small family, but the causative nature of this mutations was not certain.[3] Karkkainen *et al.* conducted a detailed *in vitro* molecular analysis of the Leu1044Pro mutation and four other VEGFR3 mutations identified in lymphedema families.[6] These mutations alter conserved amino acids in tyrosine kinase domains of VEGFR3 (FIG. 1). When these mutations are transiently overexpressed in 293T human embryonic kidney cells, which do not normally express VEGFR3, the mutant receptors do not display the ligand-independent autophosphorylation observed in 293T cells overexpressing the wild-type receptor. Stable expression of the mutant receptor confirmed that cells expressing mutant receptors failed to show ligand-dependent autophosphorylation in response to VEGFC and were less effective than wild-type receptors in eliciting downstream signaling *in vitro*. Co-expression of mutant and wild-type receptor alleles showed reduced phosphorylation of the mutant receptor but indicated that the mutant receptor was capable of forming heterodimers with the wild-type. Pulse-chase experiments using metabolically labeled receptors revealed that the mutant receptor was more stable than the wild-type, having an approximately 40% longer half-life and suggesting that the mutant receptor is internalized and degraded less efficiently than wild-type. These molecular findings were confirmed in an *in vitro* analysis of the His1035Arg mutation in kinase domain II reported by Irrthum *et al.*[7]

VEGFR3 is a member of the platelet-derived growth factor receptor subfamily of class III receptor tyrosine kinases.[8,9] VEGFR3 and the related family members VEGFR1 and VEGFR2 are expressed almost exclusively on endothelial cells and mediate the actions of the vascular endothelial growth factor family members VEGFA, VEGFB, VEGFC, and VEGFD. VEGFC

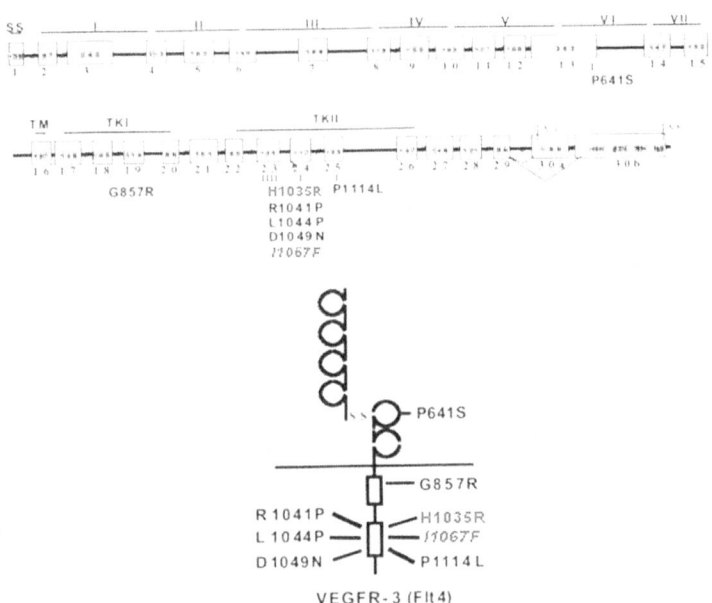

FIGURE 1. Genomic organization of the VEGFR3 gene and location of mutations causing Milroy disease. The I1067F mutation is responsible for the mouse *Chy* phenotype. The P641S mutation is a population polymorphism. The remaining mutations were identified in human lymphedema families.

and VEGFD bind to VEGFR3 and activate downstream signaling.[10,11] The VEGFRs participate in several processes essential in angiogenesis and lymphangiogenesis, including endothelial cell migration, proliferation, and survival. VEGFR3 appears to be the most critical of these receptors in the context of normal and pathological lymphatic development.

In the mouse, disruption of members of the VEGFR family leads to early embryonic death due to a failure of blood vessel development. VEGFR3-deficient embryos appear grossly normal through embryonic day 9 (E9.0) with normal-appearing myocardium, endocardium, and small blood vessels. However, defects are observed in sites of VEGFR3 expression, with irregularly formed large vessels with defective lumens. By E10, embryos are severely growth retarded with underdeveloped yolk sac vasculature. VEGFR3 appears to be necessary for normal remodeling of the yolk sac capillary network into complex vitelline vessels.[12] In normal mice, by embryonic day 13 and in subsequent development, VEGFR3 expression is largely restricted to the lymphatic endothelium,[13,14] and ablation of VEGFR3 signaling at this stage of development leads to aplasia of the lymphatics.[15] Heterozygous embryos and newborns are phenotypically normal, with normal-appearing lymphatics. Because deficiency of VEGFR3 is developmentally lethal prior to the emergence of the lymphatic vessels, a more informative model for altered lymphangiogenesis is the targeted dermal expression of soluble VEGFR3. Expression of a chimeric protein consisting of the extracellular, ligand-binding domain of VEGFR3 and the Fc domain of the immunoglobulin-gamma chain under the control of the keratin 14 promoter, leads to high-level expression in the skin around E14.5 and neutralizes the activity of VEGFC and VEGFD. This disruption of VEGFR3 signaling inhibits the formation of dermal lymphatics. Expression of the VEGFR3-Ig transgene also leads to the regression of previously formed lymphatics, indicating that intact signaling through VEGFR3 is vital to both the formation and maintenance of the normal lymphatic vasculature.[15] The gross phenotype of the VEGFR3-Ig mouse resembles that of Milroy's congenital lymphedema with hypoplasia or aplasia of the dermal lymphatics with impaired macromolecular transport, and increased fluid accumulation with visible swelling of the extremities. Older mice show thickening of the skin and dermal fibrosis.

Congenital lymphedema, characterized by aplasia or hyperplasia of the lymphatics, has been described in dog,[16,17] pig,[18] and cattle,[19] but none of these animal models has been characterized at the molecular level. The *Chy* mouse was first described by Lyon and Glenister as an ethyl nitrosourea-induced mutation characterized by a gross phenotype of chylous ascites and swollen, edematous hind feet.[20,21] The *Chy* phenotype maps to mouse chromosome 11 in a region homologous to human chromosome 5q. Like humans with Milroy's disease, the *Chy* mouse has an inactivating mutation in the tyrosine kinase II domain of VEGFR3 and hypoplastic cutaneous lymphatic vessels.[22] When the *Chy* mouse was mated with the heterozygous VEGFR3 null mouse, no viable *Chy*/– offspring were born. At E10.5, *Chy*/– embryos were growth retarded, similar to VEGFR3 null embryos.

The local delivery of VEGFC, a ligand for VEGFR3, by adeno-associated virus-mediated delivery of recombinant human VEGFC stimulated the production of functional lymphatic vessels in the ears of *Chy* mice. In addition, crosses between the *Chy* mouse and the K14-VEGF-C156S mouse, which overexpresses the VEGFR3-specific ligand VEGF-C156S in skin keratinocytes,[23] led to restored lymphatic function in the *Chy* × K14-VEGF-C156S offspring. These experiments demonstrate that normal lymphatic function can be restored in the VEGFR3-mutant *Chy* mouse by delivery of an excess of receptor ligand.

Milroy's congenital lymphedema due to VEGFR3 mutation is inherited as an autosomal dominant phenotype. Mice heterozygous for a null allele at the VEGFR3 locus display no overt phenotype, and humans with chromosomal abnormalities leading to deletion of the chromosome region containing the

VEGFR3 gene do not display lymphedema as a phenotype.[24,25] This suggests that lymphedema is not due to haploinsufficiency for VEGFR3.

Endocytosis of activated tyrosine kinase receptors through clathrin-coated pits is a well-established mechanism for regulating the duration of receptor-mediated signaling.[26] The observation that mutant VEGFR3s have a half-life significantly longer than wild-type and accumulate on the cell surface when co-expressed with wild-type, suggests a mechanism for the dominant pattern of expression of the lymphedema phenotype in families with VEGFR3 mutations. It appears that signaling through other members of the VEGFR family and/or residual signaling through wild-type and mutant heterodimeric receptors is sufficient to support normal vascular development in heterozygous carriers of kinase-negative VEGFR3 receptor mutations. However, reduced signaling through normal and mutant heterodimeric receptors coupled with an increased stability of mutant receptors is not sufficient to support normal development and maintenance of lymphatics in heterozygous carriers. The documented variation in appearance of the lymphatics in congenital lymphedema, from mild hypoplasia to aplasia, may reflect varying levels of residual kinase function and receptor stability in individuals with different VEGFR3 mutations.

Families with mutations in the VEGFR3 are characterized by an early age-at-onset, with 96% of cases being congenital. Lymphedema is the predominant phenotypic manifestation in families with documented VEGFR3 mutations. The overall penetrance in these families is 0.62, with sex-specific penetrances of 0.53 and 0.71 in males and females, respectively. The female-to-male ratio of 1.5 estimated from families with VEGFR3 mutations is similar to that reported by Smeltzer et al.,[27] but not as skewed as that estimated from earlier phenotype-based studies.[28–30] Neither the reduced penetrance nor the skewed sex ratio among affected individuals in families with molecularly defined mutations is understood.

LYMPHEDEMA–DISTICHIASIS (McKUSICK #153100)

Lymphedema–distichiasis (LD) is an autosomal dominant syndrome with the onset of lymphedema at or around puberty. Most affected individuals have distichiasis, an extra row of eyelashes arising inappropriately from the meibomian glands, and lymphedema, characterized by bilateral hyperplasia of the lymphatics.[31] LD is associated with other congenital anomalies, including congenital heart disease (16%) and, less frequently, pterygium, ptosis, cleft lip/cleft palate, and venous malformations.[32–37]

Mangion et al. first mapped a locus for LD to chromosome 16q24.3,[38] and this interval was narrowed somewhat by Bell et al.[39] Fang et al. identified mutations in the FOXC2 gene in two families, segregating for lymphedema-distichiasis.[40] The role of FOXC2 mutation in LD was quickly confirmed by several groups,[41–43] and more than 30 mutations have been characterized. All except one of the LD mutations in FOXC2 is predicted to lead to a truncation of the protein, either through a nonsense mutation or through a small insertion or deletion leading to downstream truncation. These mutations are summarized in FIGURE 2. A single missense mutation leading to a serine to leucine change in the forkhead domain of the protein was reported by Bell *et al.*[43] There is no apparent clustering of either the site of the mutation or the predicted protein truncation point among families. The site of the mutations does not correlate with the reported phenotypic features in affected families.

FOXC2 is a member of the forkhead family of transcription factors, the first of which was identified by Weigel *et al.*[44] while studying the Drosophila forkhead mutation that causes the homeotic transformation replacing ectodermal portions of the gut by entopic head structures. For structural reasons, members of this transcription factor family are also referred to as winged helix proteins and participate in a wide variety of developmental processes dur-

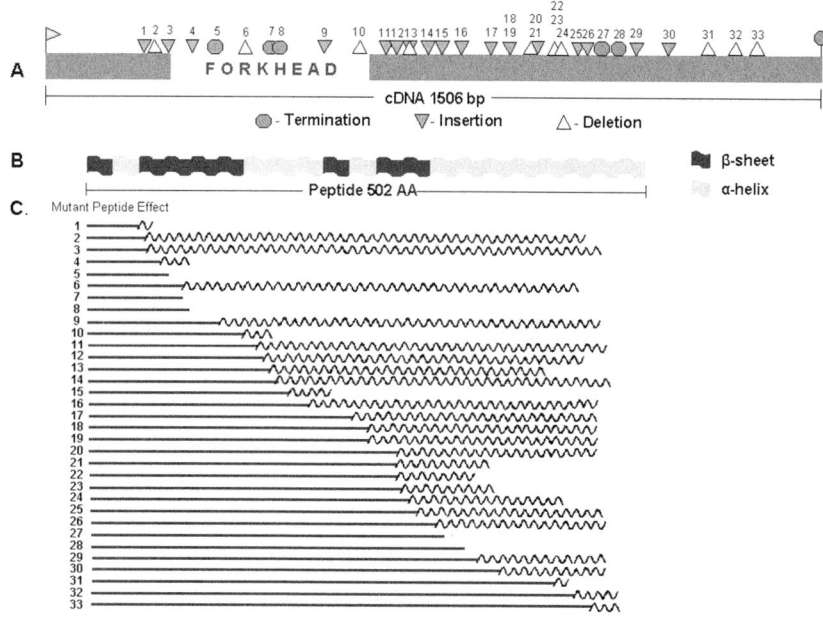

FIGURE 2. Schematic of the FOXC2 gene and mutations causing lymphedema–distichiasis syndrome. *Straight lines* indicate the normal peptide product predicted from these mutations, and *uneven lines* indicate the predicted missense portion of the peptide caused by insertion or deletion mutations.

ing embryogenesis. Mice homozygous for a null mutation in FOXC2 die prenatally (~E13.5) or perinatally and have a lethal heart defect and skeletal abnormalities.[45,46] Heterozygous wt/-FOXC2 mice were initially reported to be normal,[46] but Smith et al.[47] reported that the heterozygous null animals have multiple anterior segment abnormalities. Unfortunately, the lymphatic system was not analyzed in either the homozygous or heterozygous null animals. Specific overexpression of FOXC2 in mouse adipose tissue led to alterations in adipocyte differentiation, metabolism, insulin sensitivity, and intracellular signaling,[48] but the lymphatics have not been characterized in these animals. Thus far, studies of genetically modified FOXC2 mice have not identified a clear role of this gene in lymphatic development.

Humans carrying truncation mutations in FOXC2 generally have distichiasis at birth and lymphedema with a peripubertal onset, although both neonatal and adult onset have been observed in some individuals.[41,42] As with lymphedema in general, female mutation carriers are more likely to express lymphedema than males, although the sex ratio does not show the striking female bias generally reported for lymphedema. Expression of other features of the lymphedema distichiasis syndrome (cleft palate, ptosis, heart defects, ectropion, cystic hygroma) is variable, both in frequency of occurrence and severity. An examination of the location of FOXC2 mutations and the predicted impact of these mutations on the structure of the protein does not reveal any pattern of correlation with the occurrence of specific phenotypic features or their severity. The lack of correlation between the position of the protein truncation in these mutations and the resulting phenotype suggests that FOXC2 is a dosage-sensitive gene and the LD phenotype is due to haploinsufficiency.

LYMPHEDEMA AND PTOSIS (McKUSICK #153000) AND YELLOW NAIL SYNDROME (McKUSICK #153300)

These two rare autosomal conditions are based on clinical descriptions in a limited number of case/family reports. The finding of lymphedema occurring with ptosis and/or yellow nails in families with documented FOXC2 mutations suggests the possibility that these two entities are variants of the lymphedema distichiasis syndrome.[42,43]

NOONAN SYNDROME (McKUSICK #163950)

Noonan syndrome (NS) is a dominantly inherited syndrome of multiple congenital anomalies including short stature, webbed neck, facial abnormalities, congenital heart defects, and mental retardation. Lymphedema is a variable component of Noonan syndrome that may be diagnosed from the perinatal period through adulthood. Witt et al. reviewed the occurrence of lymphedema in NS and suggested that rather than being a nonspecific consequence of a genetic defect, edema due to lymphatic vessel dysplasia might play a causative role in the Noonan phenotype by disrupting cell migration or organ placement during development.[49] The NS gene was mapped to chromosome 12 by Jamieson et al.,[50] and the genetic interval was further narrowed by Brady et al.[51] and Legius et al.[52] Tartaglia et al.[53] identified mutations in the protein tyrosine phosphatase SHP-2 (PTPN11) gene as the cause of NS in 13 families, accounting for approximately 50% of the NS families in their sample. PTPN11, which encodes the nonreceptor protein tyrosine phosphatase SHP-2, participates in signaling cascades activated by a number of growth factors, cytokines, and hormones,[54–56] and might participate in regulating downstream signals during lymphangiogenesis. Direct sequencing of PTPN11 in primary lymphedema probands failed to detect causative mutations in the gene.[57] Thus, the relationship between lymphedema occurring in NS and primary lymphedema is unresolved.

AAGENAES' SYNDROME (McKUSICK #214900) AND HENNEKAM LYMPHANGIECTASIA-LYMPHEDEMA (#235510)

These two well-characterized autosomal recessive syndromes include lymphedema as a prominent feature. Aagenaes' syndrome[58–60] is a complex syndrome characterized by severe neonatal cholestasis with hyperbilirubinemia, increased serum concentrations of bile acids, giant cell transformation of hepatocytes, and lymphedema. Lymphedema, with hypoplastic lymphatic vessels, is manifest at birth or in childhood and is a severe, chronic condition. Bull et al.[61] mapped a gene for Aagenaes' syndrome to a 6.6-cM region of chromosome 15q in Norwegian families, but this region contains numerous genes, none of which are obvious candidate genes for lymphatic development.

Hennekam's syndrome is characterized by lymphedema of the lower extremities, facial anomalies, mental retardation and lymphangiectasia.[62] The genetic basis of the Hennekam syndrome is not known.

SUMMARY

To date, family studies followed by positional candidate gene analysis have provided the major clues to the etiology of primary lymphedema. Given the profound clinical and genetic heterogeneity in lymphedema, the study of large families segregating for lymphedema phenotypes offers one of the most promising avenues for dissecting the molecular basis of lymphedema and extensive normal lymphatic development. In addition, the identification of the VEGFR3 pathway as a cause of lymphedema has provided the first clues to developing a rational therapeutic approach to the treatment for more commonly occurring secondary lymphedema.

REFERENCES

1. MILROY, W.F. 1892. An undescribed variety of hereditary oedema. N.Y. Med. J. **56:** 505–508.
2. NONNE, M. 1891. Vier fälle von elephantiasis congenita heredituria. Virchows Arch. Pathol. Anat. **125:** 189–196.
3. FERRELL, R.E., K.M. LEVINSON, J.H. ESMAN, et al. 1998. Hereditary lymphedema: evidence for linkage and genetic heterogeneity. Hum. Mol. Genet. **7:** 2073–2078.
4. WITTE, M.H., R. ERICKSON, M. BERNAS, et al. 1998. Phenotypic and genotypic heterogeneity in familial Milroy lymphedema. Lymphology **31:** 145–155.
5. EVANS, A.L., G. BRICE, V. SOTIROVA, et al. 1999. Mapping of primary congenital lymphedema to the 5q35.3 region. Am. J. Hum. Genet. **64:** 547–555.
6. KARKKAINEN, M.J., R.E. FERRELL, E.C. LAWRENCE, et al. 2000. Missense mutations interfere with VEGFR-3 signalling in primary lymphedema. Nature Genet. **25:** 153–159.
7. IRRTHUM, A., M.J. KARKKAINEN, K. DIVRIENDT, et al. 2000. Congenital hereditary lymphedema caused by a mutation that inactivates VEGFR3 tyrosine kinase. Am. J. Hum. Genet. **67:** 295–301.
8. VAN DER GEER, P., T. HUNTER & R.A. LINDBERG. 1994. Receptor protein-tyrosine kinases and their signal transduction pathways. Annu. Rev. Cell Biol. **10:** 251–337.
9. KLAGSBRUN, M. & P.A. D'AMORE. 1996. Vascular endothelial growth factor and its receptor. Cytokine Growth Factor Rev. **7:** 259–270.
10. JOUKOV, V., K. PAJUSOLA, A. KAIPAINEN, et al. 1996. A novel vascular endothelial growth factor, VEGFC, is a ligand for the Flt4 (VEGFR3) and KDR (VEGFR2) receptor tyrosine kinases. EMBO J. **15:** 290–298.
11. ACHEN, M.G., M. JELTSCH, E. KUKK, et al. 1998. Vascular endothelial growth factor D (VEGFD) is a ligand for the tyrosine kinases VEGF receptor 2 (Flk1) and VEGF receptor 3 (Flt4). Proc. Natl. Acad. Sci. USA **95:** 548–553.
12. DUMONT, D.J., L. JUSSILA, J. TAIPALE, et al. 1998. Cardiovascular failure in mouse embryos deficient in VEGF receptor-3. Science **282:** 946–949.
13. KAIPAINEN, A., J. KORHONEN, T. MUSTONEN, et al. 1995. Expression of the fms-like tyrosine kinase-4 gene becomes restricted to lymphatic endothelium during development. Proc. Natl. Acad. Sci. USA **92:** 3566–3570.
14. KUKK, E., A. LYMBOUSSAKI, S. TAIRA, et al. 1996. VEGF-C receptor binding and pattern of expression with VEGFR-3 suggests a role in lymphatic vascular development. Development **122:** 3829–3837.

15. MAKINEN, T., L. JUSSILA, T. VEIKKOLA, et al. 2001. Inhibition of lymphangiogenesis with resulting lymphedema in transgenic mice expressing soluble VEGF-receptor-3. Nature Med. **7:** 199–205.
16. PATTERSON, D.F., W. MEDWAY, H. LUGINBUHL & S. CHACKO. 1967. Congenital hereditary lymphedema in the dog. Clinical and genetic studies. J. Med. Genet. **4:** 145–152.
17. LUGINBUHL, H., S.K. CHACKO, D.F. PATTERSON & W. MEDWAY. 1967. Congenital hereditary lymphedema in the dog. Pathological studies. J. Med. Genet. **4:** 153–165.
18. VAN DER PUTTE, S.C. 1978. Congenital hereditary lymphedema in the pig. Lymphology **11:** 1–9.
19. MORRIS, B., D.C. BLOOD, W.R. SIDMAN, et al. 1954. Congenital lymphatic oedema in Ayrshire calves. Austr. J. Exp. Biol. **32:** 265–274.
20. LYON, M.F. & P.H. GLENISTER. 1984. Chylous ascites (Chy). Mouse News Lett. **71:** 26.
21. LYON, M.F. & P.H. GLENISTER. 1986. Gene order of chy-vt-re on mouse chromosome 11. Mouse News Lett. **74:** 96.
22. KARKKAINEN, M.J., A. SAARISTO, L. JUSSILA, et al. 2001. A model for gene therapy of human hereditary lymphedema. Proc. Natl. Acad. Sci. USA **98:** 12677–12682.
23. VEIKKOLA, T., L. JUSSILA, T. MAKINEN, et al. 2001. Signaling via vascular endothelial growth factor receptor-3 is sufficient for lymphangiogenesis in transgenic mice. EMBO J. **20:** 1223–1231.
24. BARBER, J.C., I.K. TEMPLE, P.L. CAMPBELL, et al. 1996. Unbalanced translocation in a mother and her son in one of two 5;10 translocation families. Am. J. Med. Genet. **62:** 84–90.
25. GROEN, S.E., K.G. DREWES, E.G. DE BOER, et al. 1998. Repeated unbalanced offspring due to a familial translocation involving chromosomes 5 and 6. Am. J. Med. Genet. **80:** 448–453.
26. DI FIORE, P.P. & P. DE CAMILLI. 2001. Endocytosis and signaling. an inseparable partnership. Cell **106:** 1–4.
27. SMELTZER, D.M., G.B. STICKLER & A. SCHIRGER. 1985. Primary lymphedema in children and adolescents: A follow-up study and review. Pediatrics **76:** 206–218.
28. SCHIRGER, A., E.G. HARRISON, JR. & J.M. JAMES. 1962. Idiopathic lymphedema: Review of 131 cases. J. Am. Med. Assoc. **182:** 14–22.
29. KINMONTH, J.B., G.W. TAYLOR & G.D. TRACY. 1957. Primary lymphedema: Clinical and lymphangiographic studies of a series of 107 patients in which the lower limbs were affected. Br. J. Surg. **45:** 1–10.
30. FEINS, N.R., R. RUBIN, T. CRAIS & J.F. O'CONNOR. 1977. Surgical management of thirty-nine children with lymphedema. J. Pediatr. Surg. **12:** 471–476.
31. DALE, R.F. 1987. Primary lymphoedema when found with distichiasis is of the type defined as bilateral hyperplasia by lymphography. J. Med. Genet. **24:** 170–171.
32. FALLS, H.F. & E.D. KERTESZ. 1964. A new syndrome combining pterygium colli with developmental abnormalities of the eyelids and lymphatics of the lower limbs. Trans. Am. Ophthalmic Soc. **62:** 248–275.
33. CHYNN, K.Y. 1967. Congenital spinal extradural cyst in two siblings. Am. J. Roentgen **101:** 204–215.

34. PAP, Z., T. BIRO, L. SZABO & Z. PAPP. 1980. Syndrome of lymphoedema and distichiasis. Hum. Genet. **53:** 309–310.
35. CORBETT, C.R.R., R.F. DALE, D.J. COLTART & J.B. KINMOUTH. 1982. Congenital heart disease in patients with primary lymphedemas. Lymphology **15:** 85–90.
36. GOLDSTEIN, S., Q.H. QAZI, J. FITZGERALD, *et al.* 1985. Distichiasis, congenital heart defects and mixed peripheral vascular anomalies. Am. J. Med. Genet. **20:** 283–294.
37. BARTLEY, G.B. & I.T. JACKSON. 1989. Distichiasis and cleft palate. Plastic Reconstr. Surg. **84:** 129–132.
38. MANGION, J., N. RAHMAN, S. MONSOUR, *et al.* 1999. A gene for lymphedema-distichiasis maps to 16q24.3. Am. J. Hum. Genet. **65:** 427–432.
39. BELL, R., G. BRICE, A.H. CHILD, *et al.* 2000. Reduction of the genetic interval of lymphoedema-distichiasis to below 2-Mb. J. Med. Genet. **37:** 725–726.
40. FANG, J., S.L. DAGENAIS, R.P. ERICKSON, *et al.* 2000. Mutations in FOXC2, a forkhead family transcription factor, are responsible for hereditary lymphedema-distichiasis syndrome. Am. J. Hum. Genet. **67:** 1382–1388.
41. FINEGOLD, D.N., M.A. KIMAK, E.C. LAWRENCE, *et al.* 2001. Truncating mutation in FOXC2 cause multiple lymphedema syndromes. Hum. Mol. Genet. **10:** 1185–1189.
42. ERICKSON, R.P., S.L. DAGENAIS, M.S. CAULDER, *et al.* 2001. Clinical heterogeneity in lymphedema-distichiasis with FOXC2 truncating mutations. J. Med. Genet. **38:** 761–766.
43. BELL, R., G. BRICE, A.H. CHILD, *et al.* 2001. Analysis of lymphedema-distichiasis families for FOXC2 mutations reveals small insertions and deletions throughout the gene. Hum. Genet. **108:** 546–551.
44. WEIGEL, D., G. JURGENS, F. KUTTNER, *et al.* 1989. The homeotic gene fork head encodes a nuclear protein and is expressed in the terminal regions of the Drosophila embryo. Cell **57:** 645–658.
45. WINNIER, G.E., L. HARGETT & B.L.M. HOGAN. 1997. The winged helix transcription factor MFH1 is required for proliferation and patterning of paraxial mesoderm in the mouse embryo. Genes Development **11:** 926–940.
46. IIDA, K., H. KOSEKI, H. KAKINUMA, *et al.* 1997. Essential role of the winged helix transcription factor MFH-1 in aortic arch patterning and skeletogenesis. Development **124:** 4627–4638.
47. SMITH, R.S., A. ZABALETA, T. KUME, *et al.* 2000. Haploinsufficiency of the transcription factors FOXC1 and FOXC2 results in aberrant ocular development. Hum. Mol. Genet. **9:** 1021–1032.
48. CEDERBERG, A., L.M. GRONNING, B. AHREN, *et al.* 2001. FOXC2 is a winged helix gene that counteracts obesity, hypertriglyceridemia, and diet-induced insulin resistance. Cell **106:** 563–573.
49. WITT, D.R., H.E. HOYME, J. ZONANA, *et al.* 1987. Lymphedema in Noonan syndrome: Clues to pathogenesis and prenatal diagnosis and review of the literature. Am. J. Med. Genet. **27:** 841–856.
50. JAMIESON, C.R., I. VAN DER BURGT, A.F. BRADY, *et al.* 1994. Mapping a gene for Noonan syndrome to the long arm of chromosome 12. Nature Genet. **8:** 357–360.
51. BRADY, A.F., C.R. JAMIESON, I. VAN DER BURGT, *et al.* 1997. Further delineation of the critical region for Noonan syndrome on the long arm of chromosome 12. Eur. J. Hum. Genet. **5:** 336–337.

52. LEGIUS, E., E. SCHOLLEN, G. MATTHIJS & J.P. FRYNS. 1998. Fine mapping of Noonan/cardio-facial cutaneous syndrome in a larger family. Eur. J. Hum. Genet. **6:** 32–37.
53. TARTAGLIA, M., E.L. MEHLER, R. GOLDBERG, *et al.* 2001. Mutations in PTPN11, encoding the protein tyrosine phosphatase SPH-2, cause Noonan syndrome. Nature Genet. **29:** 465–468.
54. FENG, G.S. 1999. SPH-2 tyrosine phosphatase: signaling one cell or many. Exp. Cell Res. **253:** 47–54.
55. STEIN-GERLACH, M., C. WALLASCH & A. ULLRICH. 1998. SPH-2 SH-2 containing protein tyrosine phosphatase-2. Int. J. Biochem. Cell Biol. **30:** 559–566.
56. TAMIR, I., J.M. DALPORTO & J.C. CAMBIER. 2000. Cytoplasmic protein tyrosine phosphatase SHP-1 and SHR2: regulators of B cell signal transduction. Curr. Opin. Immunol. **12:** 307–315.
57. FERRELL, R.E. & D.N. FINEGOLD. 2002. Personal communication.
58. AAGENAES, O., C.B. VAN DER HAGEN & S. REFSUM. 1968. Hereditary recurrent intrahepatic cholestasis from birth. Arch. Dis. Child **43:** 646–657.
59. AAGENAES, O., H. SIGSTAD & R. BJORN-HANSEN. 1970. Lymphedema in hereditary recurrent cholestasis from birth. Arch. Dis. Child **45:** 690–695.
60. AAGENAES, O. 1998. Hereditary cholestasis with lymphedema. Scand. J. Gastroenterol. **33:** 335–345.
61. BULL, L.N., E. ROCHE, E.J. SONG, *et al.* 2000. Mapping of the locus for cholestasis-lymphedema syndrome (Aagenaes syndrome) to a 6.6 cM interval on chromosome 15q. Am. J. Hum. Genet. **67:** 994–999.
62. HENNEKAM, R.C.M., R.A. GEERDINK, B.C.J. HAMEL, *et al.* 1989. Autosomal recessive intestinal lymphangiectasia and lymphedema, with facial anomalies and mental retardation. Am. J. Med. Genet. **34:** 593–600.

Contractility Patterns of Normal and Pathologically Changed Human Lymphatics

WALDEMAR L. OLSZEWSKI

Department of Surgical Research and Transplantation, Medical Research Center, Polish Academy of Sciences, Warsaw, Poland

ABSTRACT: Human leg lymphatics contract spontaneously, rhythmically propelling lymph. This intrinsic property regulates the fluid environment in the intercellular space in skin, connective tissue, and perivascular spaces. The pressures generated by lymphatic contractions constitute the main force for lymph flow. This mechanism is of utmost importance during night rest, anesthesia, and immobilization, as well as in those with damaged peripheral motor neurons. All supporting forces are only secondary to those created by spontaneous lymphatic contractions. The intrinsically regulated lymph flow depends on tissue fluid and lymph production rate. The transport capacity of lymphatics adjusts to the fluid load. Under physiological conditions, limb activity and position only slightly change lymph flow. In obstructive lymphedema, high lymph pressures can be observed. They are usually generated by leg muscle contraction, whereas the spontaneous contractility of lymphatics becomes ineffective in lymph transport because of low generated pressures and lymphatic valve insufficiency. The knowledge of lymph flow in normal and lymphedematou limbs will be useful in the derivation of rational treatments for lymphedema.

KEYWORDS: lymphatics; contractility; lymph; flow; pressure

Under normal conditions, the main propelling force for lymph flow are the rhythmic contractions of lymphangions (segments of lymphatics between two unidirectional valves) that generate lymph pressures high enough to move the intralymphatic fluid centripetally. All other factors, such as mus-

Address for correspondence: W.L. Olszewski, M.D., Ph.D., Medical Research Center, Pawinski Str. 5, 02106 Warsaw, Poland. Voice: +48226685316; fax +48226685334.
wlo@cmdik.pan.pl

cular contractions, respiratory movements, and arterial pulsations, are secondary to spontaneous contractions of lymphatics. They may, however, become important in the setting of mechanical lymph stasis, clinically diagnosed as lymphedema.[1] The question arises of how essential spontaneous lymph flow might be for the prevention of edema formation, both under physiological conditions and in cases of mechanical obstruction of lymphatics, for example, after cancer surgery (lymphadenectomy) and following chronic skin infections with secondary damage to lymphatics and lymph nodes. How important would it be to externally regulate lymph flow through lymphatics in health and disease? Would the forced lymph flow promote lymphangiogenesis in patients with local obstruction of lymphatics? Should we search for drugs that increase the contracting forces of lymphatics? Or, should we concentrate only on drugs that prevent the development and progression of degenerative changes in lymphatics during inflammation and in the hypertension phase after surgical dissection of lymph nodes with afferent lymphatics? These questions will be answered only if we increase our knowledge of the mechanisms of lymph flow in health and in diseases of the lymphatic system, and how these mechanisms affect the tissues drained by lymphatics.

Although observations on spontaneous contractility of animal lymphatics go back to the seventeenth century, in humans it was not until we started our studies in 1980 on volunteers and patients with lymphedema. Kinmonth and Taylor[2] reported rhythmic contractility of the thoracic duct in man. Kinmonth and Taylor[3] also observed spontaneous rhythmic contraction of human retroperitoneal lymphatics in cases with chylous reflux. In 1963, Szegvari et al.[4] published two pictures of contracted and relaxed foot lymphatics in a patient prepared for lymphangiography. In none of these cases were lymph flow and pressures recorded. In 1968, Olszewski et al.[5] reported on the rhythmic contractility of leg, thigh, and retroperitoneal lymphatics in patients with so-called hyperplastic lymphedema. The lymphatics were first observed on cinelymphangiography and then after exposure during the operation for surgical lymphovenous shunt. The contractions appeared in a series of 6 to 10 waves, each wave lasting 5 to 8 sec. General anesthesia, muscle relaxants (Flaxedil), arterial pulse, spontaneous and artificial respiration, and apnea had no influence on the rhythm of spontaneous contractions. Spontaneous contractility of human prenodal lymph vessels has also been observed during X-ray lymphangiography. In a study by Armenio et al.,[6] 75% of technically satisfactory serial lymphangiograms recorded in human subjects revealed morphological changes in the outline of lymph vessels in lower limbs, changes most likely caused by spontaneous contractility.

The aim of this article is to summarize the contemporary knowledge on the spontaneous contractility of human leg lymphatics and the pressure/flow relationship regulated by the contracting forces.

PRESSURE AND FLOW IN HUMAN LEG LYMPHATICS

There are considerable differences in reported intralymphatic pressures. The differences depend on whether lateral or end pressures have been measured, whether the systolic or the diastolic phase of the contraction of the lymphatic segment has been recorded, and whether the muscles were actively contracting while measurements were taken.

The pressures in the prenodal lymphatics are similar, irrespective of the drained areas. They depend on the intrinsic contractility of lymphatics,[7] on vascular resistance, and partly on active and passive movements of neighboring muscles or of the muscles of the drained extremity. The first studies of spontaneous rhythmic contractility of prenodal lymphatics (FIG. 1) in man, with intravascular pressure and lymph flow recordings, were performed by Olszewski and Engeset.[1,7,8] To study the efficiency of intrinsic contractions of the lymphatics on lymph propulsion in human legs, the end and lateral pressures, as well as lymph flow, were measured in leg subcutaneous lymph vessels. Systolic lymph end pressures generated by intrinsic contractions of

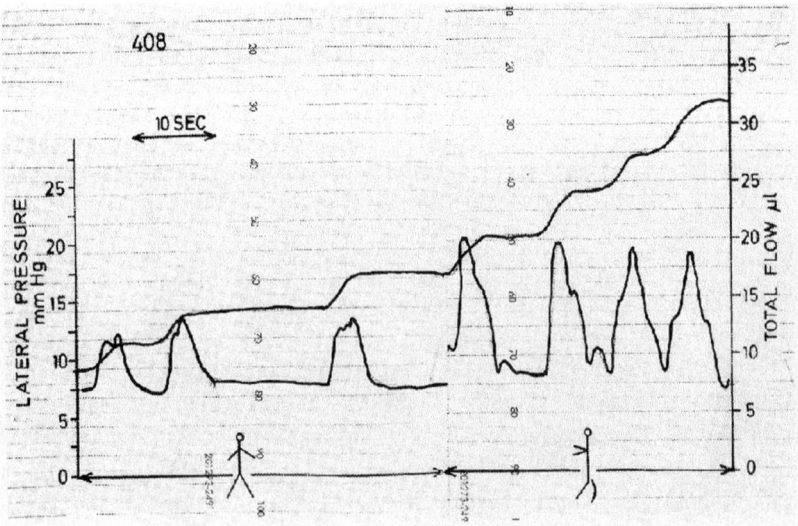

FIGURE 1. Lymph pressure and flow recordings in a superficial leg lymphatic of an individual in upright position. There is no hydrostatic component of the lymph pressure. In normal legs the diastolic pressure is only 7 mmHg. Tiptoeing increases the frequency and amplitude of lymphatic pulses as well as lymph flow (*upper curve*). Note that each pulse has a different shape, which is most likely caused by superimposition on the main wave of low-amplitude pulses from distal lymphangions.

the lymphatics ranged between 12 and 70 mmHg, but in some cases reached values above 100 mmHg. Systolic lymph lateral pressures with free lymph flow were lower and ranged between 5 and 30 mmHg.

Simultaneous measurements of lymph pressures and flow have been introduced when we constructed a low-flow flowmeter, with an accuracy of 0.1 µL within the flow range between 0.1 and 6 µL/sec.[9]

A subcutaneous lymph vessel of the leg was cannulated, against the direction of lymph flow, according to the techniques described previously.[10,11] The lymph vessel drained lymph from the skin, subcutaneous tissue, and perimuscular fascia of the foot and anterior aspect of the lower leg.

End Intralymphatic Pressure

Horizontal Position—No Movements of Leg or Foot

The end pressures hovered near 0 mmHg in the first few seconds after closing the stopcock. When the pressure reached a level of 5 to 25 mmHg, rhythmic pulsations appeared, composed of groups of waves seen at intervals of 30 sec to over 1 min. In some cases, pulsations were observed for at least 20 to 30 min, followed by short relaxation intervals. The systolic pressures reached a mean level of 37.9 mmHg. In some cases, they reached levels of approximately 50 mmHg; in some others, as high as 120 mmHg. Cases were also observed in which the diastolic pressure remained stable slightly above 0 mmHg, only to increase to 30 to 40 mmHg during the pulsation waves. The amplitude of pulse waves ranged from 3 to 35 mmHg, with a mean of 14.36 mmHg. The mean pulse frequency was 5.35 per minute. The duration of single pulse waves was on the average 6.0 ± 0.32 sec.

Horizontal Position—Rhythmic Flexing of the Foot

Rhythmic flexing of the foot with sole pressure against the bed board at a rate of 30 per minute did not increase the mean systolic pressure above the values from the rest period, but brought about an increase in lymphatic pulse frequency ($P < 0.025$), with a moderate decrease in pulse amplitude ($P < 0.05$). Voluntary muscular contractions of the leg produced pressure amplitudes as low as 0.5 to 3 mmHg. They could be seen superimposed on the spontaneous lymphatic pulse waves.

Change from the Horizontal to Upright Position

In some cases, the mean end pressure after the subjects assumed an upright position was slightly higher than that recorded with a horizontal position. There was an increase in pulse frequency, and there were shorter intervals between the pulsations.

Upright Position—No Movements of Foot and Leg

The systolic end pressures reached different individual maximum values as quickly as a few minutes after the recording was started, ranging from 20 to more than 120 mmHg with a mean of 44.7 mmHg. There were no statistically significant differences in mean pressure with the period of rest in a horizontal position. The frequency of lymphatic pulsation was 7.55 per minute. It was slightly higher than while maintaining a resting horizontal position ($P < 0.05$). The pulse wave amplitude was, on average, 9.45 mmHg, with a range of 4 to 25 mmHg. It was lower than in the resting horizontal position ($P < 0.01$).

Upright Position—Rhythmic Rising on Toes

Rising on the toes, thus contracting the foot and calf muscles, did not produce any rise in the mean systolic end pressures over the values from the rest period. However, there was an increase in pulse frequency ($P < 0.05$). No change in pulse amplitude was observed. The pressure waves produced by calf muscle contractions had an amplitude of not more than 3 mmHg.

A great diversity of pressure patterns was found in the investigated individuals with respect to the shape of the pulse curve, the levels of diastolic and systolic pressure, and the time to reach maximum lymph end pressure. These differences may depend on the size of the vessel intercommunicating branches, and valve competency.

Lateral Intralymphatic Pressures and Flow

Horizontal Position—Rest or Rhythmic Movements of Foot and Leg

The mean lateral systolic pressure in the lymphatics with free lymph flow was 13.5 mmHg during rest. It was significantly lower than in vessels with obstructed lymph flow ($P < 0.005$). The mean pulse amplitude was 8.8 mmHg, and the pulse frequency 2.42 per minute. The latter was lower than during measurement of end pressures ($P < 0.01$). The pressure in the periods between the pulse waves was always between zero and several mmHg.

Rhythmic active movements of the foot and leg did not significantly increase the mean systolic pressures, but there was a rise in pulse frequency to 4.4 per minute ($P < 0.01$). The systolic pressure and pulse frequency was significantly lower than while measured in vessels with obstructed lymph flow ($P < 0.01$ and $P < 0.05$, respectively).

During rest, flow occurred only simultaneously with the pulse. There was no flow in the intervals between the pulse waves. Also, during foot and leg movements, flow occurred only when the intrinsic pulse waves were recorded. The pressure waves produced by limb muscular contractions had an amplitude of 1 to 3 mmHg and were too weak to produce any lymph flow. Active movements of foot and calf increased the lymph flow from 0.3 ± 0.25 to 1.3 ± 0.9. The frequency of lymphatic pulse also increased, whereas pulse amplitude and stroke volume remained rather unchanged.

Upright Position—Rest or Rhythmic Rising on Toes

Standing was accompanied by a mean systolic lateral pressure of 15.23 mmHg and was not different from that in horizontal position. However, it was significantly lower than the end systolic pressure measured in the upright position ($P < 0.0025$). The mean pulse amplitude was 8.37 mmHg, and the pulse frequency was 3.19 per minute. They did not differ from the values obtained during a reclining position. A significant difference was found between pulse frequencies during recording of end and lateral pressures in upright position ($P < 0.05$).

Rising on toes brought about an increase in mean systolic pressure over the values from the resting period to 23.8 mmHg ($P < 0.01$) and in pulse frequency to 5.5 per minute ($P < 0.005$). Again, the pressures and pulse frequencies were lower during recording of lateral than during recording of end pressures ($P < 0.01$). The mean lymph flow during the rising-on-toes movement increased from 0.43 ± 0.3 to 1.3 ± 1.2. Pressure waves generated by muscle contractions did not affect flow. The frequency of pulse increased from 2.8 ± 2.0 to 5.5 ± 1.5. No statistically significant increase in pulse pressure and stroke volume was noted.

In order to determine which of the highest pressures generated by spontaneous contractions are still effective for propelling lymph in case of proximal resistance to flow, a series of experiments was carried out. The external tip of the lymphatic cannula was raised to various levels in order to increase hydrostatic pressure above the level of the cannulated lymphangion. Increase in outflow resistance caused increased pulse frequency by 20 to 30% with subsequent decrease in pulse amplitude, stroke volume, and flow rate. At the mean intralymphatic pressure of 34 mmHg, lymph flow ceased despite high pulsation rate.

Effect of Massage on Pulsation and Lymph Flow

Massaging the top of the foot performed at a rate of 30 compressions per minute increased pulse frequency to almost the same level as did active movement of the foot. Flow was only observed during pulse waves. There was no flow in the period between the pulses, although the massaging was continued. The mean lymph flow in the period of massaging was similar to that recorded during active movement of the foot.

Lymph Pressure and Flow in Obstructive Lymphedema

Lymph pressure and flow in leg lymphatics in obstructive lymphedema depend on factors affecting the anatomy of the lymphatic network and lymphatic vessel wall: persistence of pathological factors that damage lymphatic endothelium and the muscular fibers sustaining the inflammatory process; lymph hypertension caused by obstruction of outflow pathways including lymph node lymph channels; accumulation of lymphocytes and their products; retention of microorganisms usually transported from the site of penetration of skin to the lymph node; and accumulation of chemical tissue products. Degenerative changes in the lymphatic wall develop, including hyperplasia of endothelial cells, thickening of the subendothelial layer of collagen fibers, a decrease in number of muscular fibers, and (eventually) total occlusion of the lumen.[12]

In the initial stages after mechanical or inflammatory damage to lymphatics, deceleration of lymph flow develops. Overloading of lymphatics with an

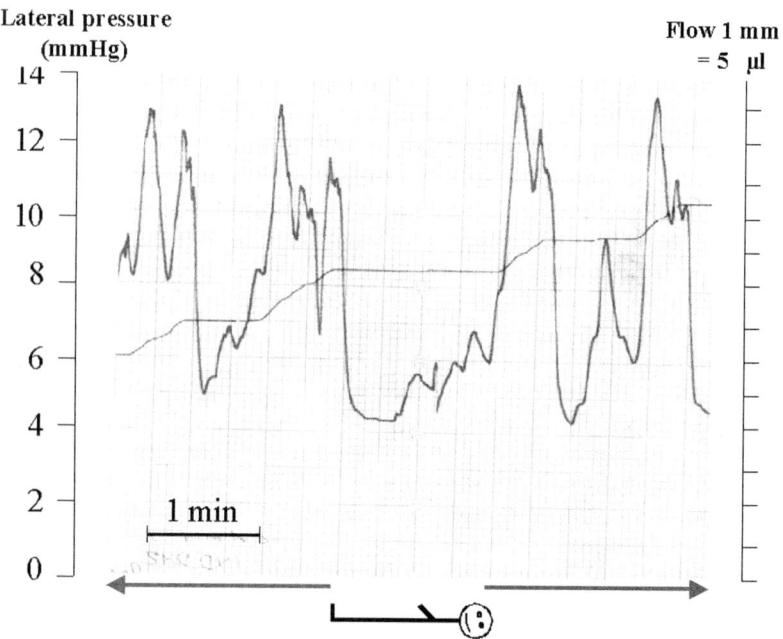

FIGURE 2. Intravascular lymph pressure recording in a leg lymphatic of a male with lymphedema. The subject is in a horizontal position. Lymphatic pulse waves come from the cannulated lymphangion with superimposed pressure waves from the more distal lymphangions. Flow (stepwise rising line) occurs only during pressures generated by the contracting lymphangion.

excess of continuously produced lymph brings about dilatation of vessels with subsequent insufficiency of unidirectional valves. Retrograde flow can be observed with external pressure exerted on proximally located tissues. The stretched lymphatics contract continuously; however, they do not generate effective pressures sufficient to propel lymph. A "to and fro" movement of lymph can be observed.

There is substantial diversity in the patterns of lymph pressure and flow observed in obstructive lymphedema. The more advanced the lymph stasis, the more irregular the recordings (FIG. 2). Interestingly, since the lymphatic outflow pathways are obstructed, muscular contractions are ineffective in evacuating excess lymph from dilated lymphatics. Consequently, the mean pressure does not decrease as would be expected (FIG. 3).

Early Stage of Lymphedema after Lymphadenectomy

In early-stage obstructive lymphedema, with dilated lymphatics, after inguinal lymphadenectomy, the pressure in a resting horizontal position is 0 or slightly above (FIG. 4). Assumption of an upright position increased diastolic

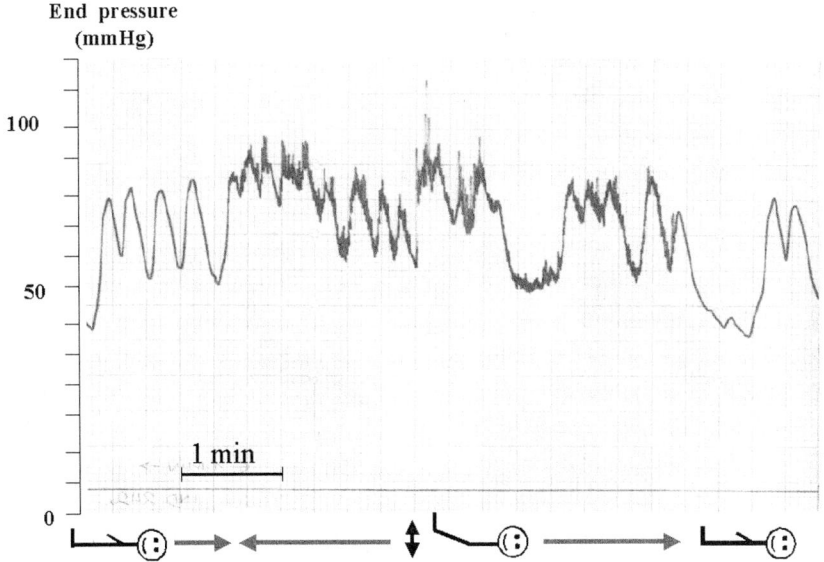

FIGURE 3. Intravascular lymph pressure recording in an externally obstructed lymphatic. Diastolic pressure reached 70 mmHg; systolic pressure, 100 mmHg. Leg muscular contractions did not significantly increase pressure, because under normal conditions muscular compartment pressures are not transmitted to the loose connective tissue where the lymphatic trunks are located.

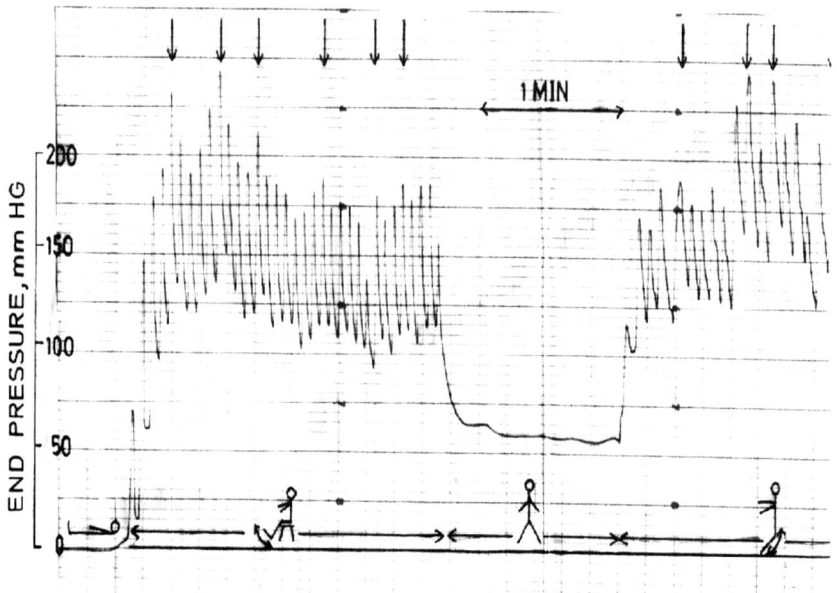

FIGURE 4. Diastolic end pressure in a resting horizontal position was 0 mmHg. When the subject assumed an upright posture, the diastolic end pressure rose to 60 mmHg (hydrostatic component, which is not observed in normal limbs). Tiptoeing elevated pressures to above 200 mmHg. The pressure waves were produced by both calf muscle contractions (not observed in normal limbs) and spontaneous contractions of lymphangions (*arrows*).

pressure to 50 or 60 mmHg (hydrostatic component, not present in normal, partially empty lymph vessels). Muscular contractions elevated end pressures to 200 mmHg and above. Both spontaneous contractions of lymphatics and voluntary contractions of calf muscles generated pressures sufficient to propel lymph. However, flow would not necessarily be in the proximal direction.

Advanced Stage of Lymphedema after Lymphadenectomy

Although the lymphatic pulse may be regular, its systolic amplitude is low, even when the outflow end becomes obstructed (end pressure). Because of insufficiency of valves, there is no lymph flow (FIG. 5).

Differences among lymph parameters in normal and lymphedematous legs have been are presented in TABLE 1.

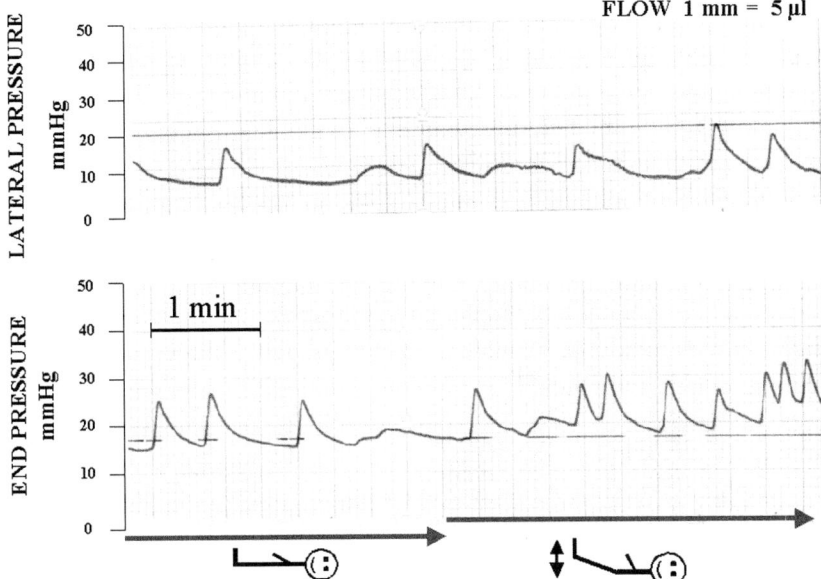

FIGURE 5. Intravascular lymphatic pressures and flow in Stage III obstructive lymphedema. In the *lower panel*, a proximal part of the lymphatic was obstructed, but pressure did not rise as high as in normal vessels and remained at the level of 20 mmHg. In the *upper panel*, with free flow, the pressure during the systolic phase (flat line) reached only 15 mmHg, not enough to produce lymph flow. Destruction of muscular fibers in the lymphatic wall presumably impaired the ability to generate effective pressure upon contraction.

TABLE 1. Differences in lymph pressure and flow in health and obstructive lymphedema

	Normal	Lymphedema
Spontaneous contractions	Present	Irregular, superimposed pressure waves from various lymphangions; low amplitude
Hydrostatic pressure component	Absent	Present
Lymphatic pulse	Approximately 6/min at rest, up to 18/min during walking	Irregular, low mean pressure, ineffective for propelling lymph. No pulse, no pressure
End-pressures during walking	Approximately 40 mmHg	Up to 200–300 mmHg
Ability of leg muscle contractions to propel lymph	Absent	Present
Ability of elastic support to improve flow	Absent	Present

DISCUSSION

Lower limb lymphatics contract spontaneously, rhythmically, and transport daily, according to our estimates, between 20 to 250 mL of lymph in a 70-kg man. Knowledge surrounding this extremely important mechanism of tissue fluid and lymph flow away from tissues, securing the intercellular physico-chemical environment, is rudimentary among clinicians and even physiologists studying the peripheral circulation. In our studies, we have found that, under physiological conditions, spontaneous contractions of segments of lymphatics generate pressures necessary for the centripetal movement of lymph. This mechanism becomes especially important for lymph transport during night rest, during anesthesia, in bed-confined or immobilized patients, and in cases of damaged peripheral motor neurons. Although important, all of the auxiliary forces, traditionally listed in textbooks, such as limb muscle contractions, respiratory movements, and pulsation of neighboring arteries, remain secondary to intrinsic contractions of lymphatics.

In cases of mechanical obstruction of lymphatic pathways or destruction of the lymphatic wall, these auxiliary forces are not sufficient to propel lymph. Consequently, edema develops, with secondary changes in the swollen tissues. In long-lasting lymph stasis, lymphatic segments still contract as they are stretched by high lymph volume in the lymphatic tree of the limb. These contractions gradually become ineffective in unidirectionally propelling lymph. The idle peristaltic movements of lymphatics can be seen during surgery or on contrast lymphography. We observe them even years following mastectomy with axillary lymphadenectomy. The highly neglected homeostatic mechanism of active tissue fluid and lymph transport from tissues to systemic circulation deserves to find its place in clinical thinking. From the point of view of future research, two basic questions arise:

1. Should we search for drugs that will specifically stimulate lymphatic contractility, in analogy to cardiac therapies?

2. What sort of treatment should be applied to eradicate the noxious factors that destroy lymphatic structure in the vessels affected by inflammation and surgical injury?

The answer to the first question is *yes*. Drugs stimulating lymphatic muscular fibers may not only increase their functional capacity, but also stimulate their proliferation. Spontaneously contracting vessels may create intraluminal pressures that would facilitate growth of lymphatics, bypassing the areas of obstruction. With respect to the second question, an immediate application of antibiotics, to mitigate the microbial activity in lymphatics and lymph nodes, as well as to protect against colonization of obstructed vessels by microorganisms, is indispensable. There are no clinical methods for the routine recognition of ongoing changes in lymphatics after trauma, infection, lymphadenectomy, and other pathological states. We do not know how the contractility of lymphatics, and subsequent lymph flow, are affected by pathological processes in the non-lymphoid tissues. The presented recordings from patients with lymphedema illustrate only a final result of the pathological process. If we were able to clinically evaluate the transport function of lymphatics in limbs and organs, we could intervene immediately and rationally.

REFERENCES

1. OLSZEWSKI, W.L. & A. ENGESET. 1980. Intrinsic contractility of prenodal lymph vessels and lymph flow in human leg. Am. J. Physiol. **239:** H775–H783.
2. KINMONTH, J.B. & G.W. TAYLOR. 1956. Spontaneous rhythmic contractility in human lymphatics. J. Physiol. (London) **133:** 3.
3. KINMONTH, J.B. & G.W. TAYLOR. 1964. Chylous reflux. Br. J. Med. **1:** 528.
4. SZEGVARI, M., A. LAKOS, F. SZONTAGH & M. FOLDI. 1963. Spontaneous contractions of lymphatic vessels in man. Lancet **1:** 1329.
5. OLSZEWSKI, W.L., S. KRUSZEWSKI & L. ZGLICZYNSKI. 1968. Obserwacje ruchow wlasnych naczyn chlonnych u chorych z obrzekiem chlonnym konczyn. Pol. Tyg. Lek. **23:** 1345–1347.
6. ARMENIO, S., F. CETTA, G. TANZINI & C. GUERCIA. 1981. Spontaneous contractility in the human lymph vessels. Lymphology **14:** 173–178.
7. OLSZEWSKI, W.L. & A. ENGESET. 1979. Lymphatic contractility. N. Engl. J. Med. **300:** 316.
8. OLSZEWSKI, W.L. & A. ENGESET. 1979. Intrinsic contractility of leg lymphatics in man — preliminary communication. Lymphology **12:** 81–84.
9. SØRENSEN, O., A. ENGESET, W.L. OLSZEWSKI & T. LINDMO. 1982. High-sensitivity optical flowmeter. Lymphology **15:** 29–31.
10. ENGESET, A., B. HARGER, A. NESHEIM & A. KOLBESTVEDT. 1973. Studies on human peripheral lymph. I. Sampling method. Lymphology **6:** 1–5.
11. OLSZEWSKI, W.L. 1977. Collection and physiological measurements of lymph and interstitial fluid in man. Lymphology **10:** 137–145.
12. OLSZEWSKI, W.L. 1977. Pathophysiological and clinical observations of obstructive lymphedema of limbs. *In* Lymphedema. L. Clodius, Ed. Thieme Verlag. Stuttgart.

Preclinical Models of Lymphatic Disease

The Potential for Growth Factor and Gene Therapy

STANLEY G. ROCKSON

Falk Cardiovascular Research Center, Stanford Center for Lymphatic and Venous Disorders, Division of Cardiovascular Medicine, Stanford University School of Medicine, Stanford, California 94305, USA

> ABSTRACT: The human disease states that are characterized by functional lymphatic insufficiency currently lack a cure. Molecular approaches may ultimately provide a therapeutic window to reverse the stigmata of both primary and secondary lymphatic insufficiency. To harness the potential therapeutic power of lymphangiogenesis, testing the safety and efficacy of the treatment response will be necessary. This, in turn, necessitates the availability of suitable preclinical animal models of the disease processes in question, along with suitable research tools to permit an assessment of the response to applied therapies. An ideal model would reproducibly and inexpensively replicate the untreated disease of human lymphedema. It would closely simulate the biology, as we understand it, of the human disease, and would replicate both the pathogenesis of the disease, including its natural history and the temporal patterns of its clinical expression. In this way, one might aspire to make valid predictions about the human applicability of therapy by extrapolation from observations in animal models. In addition to the availability of suitable animal models, the required investigative tools must also be available. In the context of lymphangiogenesis, to assess the therapeutic response, one must certainly possess the ability to recognize newly developed lymphatic vasculature. Sophisticated immunohistochemical and imaging techniques make this increasingly feasible. Initial experimental observations indicate that growth factor and gene therapy with VEGF-C holds promise for the treatment of both primary and secondary forms of lymphedema.
>
> KEYWORDS: VEGF-C; VEGFR-3; lymphedema; lymphangiogenesis; animal models

Address for correspondence: S.G. Rockson, M.D., Falk Cardiovascular Research Center, Stanford Center for Lymphatic and Venous Disorders, Division of Cardiovascular Medicine, Stanford University School of Medicine, Stanford, CA 94305.
srockson@cvmed.stanford.edu

There is a broad spectrum of human clinical pathology that leads to a distortion or loss of the lymphatic transport capacity.[1] These lymphatic disorders, at times expressed as a consequence of heritable pathology,[2] or, alternatively, acquired after surgical disruption, trauma, or infection,[3] can occasion a latent disturbance of lymphatic function and, quite commonly, lead to an overt clinical disorder of the microcirculation. These latter diseases, which collectively manifest the clinical presentation of lymphedema, are characterized by chronic edema, loss of normal cutaneous and subcutaneous tissue architecture, distortion of regional immune function, and substantial psychological[4,5] and physical[6] disability. While clinical management protocols exist, the efficacy of existing modalities is variable and, often, unsatisfactory. Lymphedema currently lacks an effective cure. For this reason, there has been substantial interest in the rapidly advancing field of growth-factor-mediated lymphangiogenesis. As a direct extension from proposed angiogenic treatment strategies for diseases of the peripheral and coronary blood vasculature,[7–9] molecular approaches may ultimately provide a therapeutic window to reverse the stigmata of both primary and secondary lymphatic insufficiency.

LYMPHANGIOGENESIS AS A BIOLOGICAL PHENOMENON

Lymphangiogenesis is a term that can be employed to designate the postnatal development of new lymphatic microvasculature by extension from the existing vessels. Recognition of this biological capacity for *de novo* lymphatic development is not a very recent event; in fact, an *in vivo* demonstration of postinflammatory lymphangiogenesis was published as early as 1937.[10] Nevertheless, despite this early anatomic identification of the biological process, it has not been until the last half-decade that a veritable explosion of information and tools has permitted this process to be harnessed for potential therapeutic applications.

It is now commonly understood that, among the mitogenic substances that initiate and regulate the growth of vascular structures, the VEGF family of molecules plays a central role.[11,12] Within this family, there are individual molecular species, namely, VEGF-C and VEGF-D, that specifically direct the development and growth of the lymphatic vasculature in the latter phases of embryonic development and in postnatal life.[13–16]

These lymphangiogenic growth factors mediate their responses by serving as ligands for the VEGFR-3 receptor on lymphatic endothelia[13]; this relatively unique receptor specificity for lymphatic structures provides an immunohistochemical tool for the reliable identification of these vessels in microscopic section.[17] Thus, in contrast to VEGF, which, when administered exogenously, mediates angiogenesis through the upregulation of VEGFR-2

> The model would reproducibly and inexpensively replicate the untreated disease
>
> It would closely simulate the biology of the human disease
>
> It would replicate the pathogenesis and natural history of the human disease

FIGURE 1. The ideal attributes of a preclinical animal model to develop and test therapeutic strategies for lymphangiogenesis.

expression, VEGF-C upregulates the VEGFR-3 receptor, leading to a lymphangiogenic response.[18] Furthermore, in transgenic mice that overexpress VEGF-C, it has been observed that the lymphatic vessels experience a hyperplastic, proliferative response, with secondary cutaneous changes.[15]

AN IDEAL ANIMAL MODEL

To harness the potential therapeutic power of these very important molecular observations, testing of the safety and efficacy of the treatment response will be necessary. This, in turn, necessitates the availability of suitable preclinical animal models of the disease processes in question, along with suitable research tools to permit an assessment of the response to applied therapies.

To attain these ends, a successful preclinical model for therapeutic lymphangiogenesis should incorporate at least several ideal attributes (FIG. 1).

The model would reproducibly and, ideally, inexpensively replicate the untreated disease of human lymphedema; it would closely simulate the biology, as we understand it, of the human disease; and would replicate the pathogenesis of the disease, including its natural history and the temporal patterns of its clinical expression. In this way, one might aspire to make valid predictions about the human applicability of therapy by extrapolation from the animal model observations.

Historically, the task of identifying such a model has been fraught with some difficulty. Dr. W.S. Halsted, who introduced the technique of radical mastectomy as a surgical approach to breast cancer cure, had an abiding interest in the phenomenon of postmastectomy lymphedema.[19] Nevertheless, his attempts to elaborate a model were unsuccessful, and this lack of success has often been replicated in the wake of his initial work.

The modern era of successful experimental lymphedema was initiated in 1968, with the work of Dr. W. Olszewski, in dogs.[20] Many of the subsequent animal models, published by a variety of investigators, represent minor modifications of the technique that he initially elaborated. The latter entails a circumferential incision at the mid-thigh of the limb, resection of the main lymphatic trunks with ligation at both ends, excision of popliteal nodes, suturing of the skin edges, and ligation of the afferent lymphatics. While successful, this is a substantial surgical intervention. The surgical trauma certainly exceeds what is undertaken in the conventional breast cancer surgeries that, despite modifications from the time of Halsted, are still quite capable of producing acquired lymphatic impairment.[21,22]

Similar approaches to that of Olszewski have been adapted to smaller laboratory animals, to capitalize on greater availability, lower cost, and telescoping of the time lapse required for clinically relevant observations.

One such model entails the creation of postsurgical lymphedema in the rat hindlimb.[23] Despite a superficial resemblance to the human state of postsurgical lymphedema, there are some inherent problems with this otherwise accessible model. Closer examination discloses an early, 30-day time frame in which there is a significant increase in limb volume. Thereafter, however, the volumes of the operated and normal limbs much more closely approximate one another, as the normal, unoperated limb also becomes edematous. This latter observation is unexplained, and does not in any manner simulate the observations in human postsurgical lymphedema. A further concern with this model is the unacceptably high death rate (15/70 treated animals).

There has been additional, more recent publication of a rodent model to simulate post-mastectomy and other cancer-associated lymphedemas.[24] Under scrutiny was the capability of surgery alone, radiation alone, or surgery and radiation, combined, to elicit a lymphedema response.

Somewhat surprisingly, the use of radiation alone or surgery alone has little capacity to produce edema: radiation alone was not successful (a divergence from the human disease state), and surgery alone produced an acute lymphedema that resolved over time (again, a divergence). It was only with combined intervention that chronic lymphedema ensued. Furthermore, in this model, there was an unacceptably high late morbidity (by analogy to human lymphedema), that included skin breakdown, occasionally extensive soft tissue fibrosis, and bone necrosis in the lymphedematous limb.

IDENTIFICATION OF LYMPHATIC VASCULATURE

In addition to the requisite availability of animal models suitable to the disease state, the required investigative tools must also be available. In the context of lymphangiogenesis, one must certainly possess the ability to recognize newly developed lymphatic vasculature to assess the therapeutic response. Historically, the recognition of lymphatic microvasculature in standard light microscopic sections has been fraught with difficulty. Previously, there has been substantial reliance upon the injection of vital dyes that, in turn, permit visual inspection of the cutaneous structures. However, more recently, additional, highly useful imaging technology has substantially enhanced this aspect of the investigative task.

One such approach entails the use fluorescent microlymphography to more accurately demonstrate these vessels in the cutaneous structures and elsewhere. This approach permits not only a qualitative description of the vasculature,[25] but also allows quantitative measures of vessel size, density, and function.[15,26–29]

Recent immunohistochemical advances have also been substantial. The utility of the VEGFR-3 expression on the lymphatic endothelia has already been discussed. However, the modest limitations inherent in the use of anti-VEGFR-3 antibodies to identify lymphatic structures underscore the importance of an, as yet, unidentified "ideal" marker for the lymphatic endothelium.[17] Ideally, such a structural component should be exclusively present or absent on either lymphatic or blood vascular structures, respectively. Furthermore, the expression patterns of this putative marker would not change during pathological processes, such as tumor development or metastasis. Ultimately, the lymphatic-specific nature of the putative marker must be verified by electron microscopy.

In addition to the VEGFR-3 receptor, a number of additional, nearly ideal markers for lymphatic endothelium have been identified in recent years. Among the most promising are LYVE-1,[30] a CD44 homologue that serves as a hyaluronan receptor on lymphatic endothelia and macrophages. Although LYVE-1 can be detected in other cells, it is generally not seen in blood vessels. PROX-1 is another marker of lymphatic vasculature in adult tissues.[31,32] An additional marker that is expressed largely, but not exclusively, by lymphatic endothelial cells is podoplanin.[33–35] It is strongly expressed in the cutaneous lymphatics. In addition to these specific cell-surface markers, a number of structural and enzymatic features of the lymphatic endothelium can facilitate its histological identification.[17] Many of the latter have been verified by ultrastructural confirmation.

CLINICAL CORRELATIONS TO HUMAN LYMPHATIC INSUFFICIENCY

The evaluation of molecular therapeutics to treat lymphatic disease presupposes that the model of choice will simulate the attributes of the human pathological state.

In a clinical context, physicians and investigators rely heavily upon radionuclide lymphoscintigraphy to identify and characterize lymphedema and its response to therapeutic intervention.[36] The diagnosis is established by the observation of interrupted transport of a subcutaneously administered, radiolabeled substance that has obligate lymphatic clearance. In addition, the identification of dermal backflow (interstitial accumulation of the radiotracer as a consequence of lymphatic insufficiency and lymph stagnation) is considered pathognomonic of lymphatic insufficiency.[37]

The consequences of chronic lymph stasis have been well studied in the human condition. These include the development of protein-rich interstitial edema, with the ultimate development of fibrosis and sclerosis of the dermal structures. In addition, local blood-flow disturbances are seen, and there is a substantial propensity to infection. The latter relates not only to regional dysfunction of the immune surveillance conferred by the lymphatic circulation, but also to the excellent microbial growth support that is afforded to saprophytic organisms by the stagnant lymph.

CURRENT MODELS FOR THE STUDY OF LYMPHANGIOGENESIS

The explosion of molecular information surrounding lymphangiogenesis has shed substantial light on the mechanisms surrounding the heritable form of lymphedema, an autosomal dominant condition known as Milroy's disease. Here, in several families, the disease has been linked to the FLT4 locus, encoding vascular VEGFR-3.[38] All disease-associated alleles analyzed had missense mutations that yielded an inactive tyrosine kinase, thereby preventing downstream gene activation. Thus, the inherited disease hampers the ability of the VEGFR-3 receptor to orchestrate lymphangiogenesis in development.

The mechanisms of the disease have been putatively delineated: it is believed that, because of an excessive stability of the mutant form of the receptor on the cell surface, the normal signaling mechanism is blunted, leading to hypoplastic development of the lymphatic vessels.[39,40]

There are model systems available in which to examine these phenomena. Quite recently, a transgenic mouse was described that has transient attributes of human primary lymphedema.[41] This transgenic animal expresses a soluble form of the VEGFR3 receptor that, in turn, elicits the expression of abnormalities in the skin, along with the transient appearance of lymphedema of the limbs. Initially, there is an apparent lack of dermal lymphatic vessel development, although, over time, there is an escape from this pattern of biological expression.

Perhaps of even greater relevance to Milroy's disease is the more recent description of the Chy mouse model.[42] This model also expresses a form of limb edema that bears a superficial resemblance to human primary lymphedema. However, in this case, the pathology is even more tightly congruent because, in the Chy mouse, the VEGFR-3 receptor is mutant, as well.

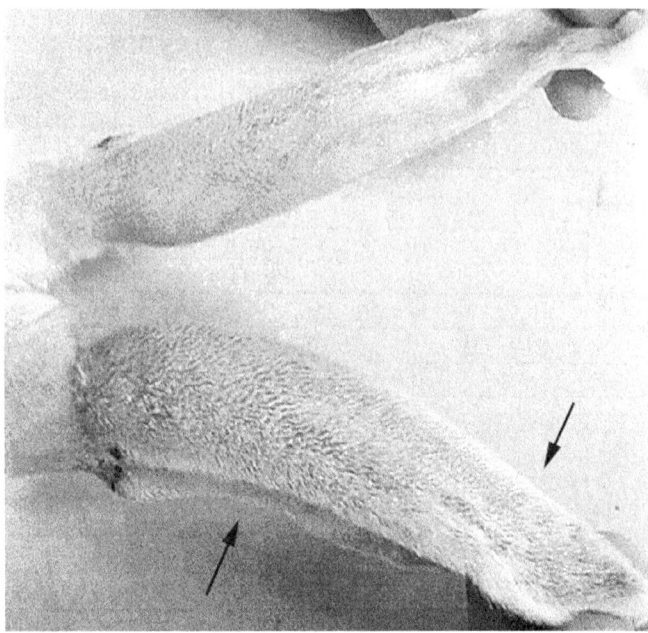

FIGURE 2. An example of the chronic, experimental, postsurgical lymphedema in the rabbit ear. The *arrows* identify the ear with lymphedema.

Therefore, it is extremely encouraging to contemplate that, in this model that so closely resembles Milroy's disease, the induction of VEGF-C overexpression through gene therapy with a viral vector, there is an observed amelioration of lymphedema that accompanies the generation of new, functional lymphatic vasculature.

In our laboratory, we have chosen to focus on the problem of acquired lymphedema and its potential responsiveness to growth-factor-mediated therapeutic lymphangiogenesis. In this endeavor, we have chosen to modify a published model of acquired, postsurgical lymphedema in the rabbit ear.[43] For this model, a segment of the skin is resected at the base of one of the ears. After identification of the major lymphatic trunks through methylene blue injection, major portions of these collecting vessels are resected, to deprive the ear of its lymphatic outflow (FIG. 2). Thus, the ear becomes the analogue of an edematous arm that becomes manifest after a surgical intervention for breast cancer. After allowance is made for healing of the surgical wound, reproducible, sustained edema of can be achieved with high tolerability and low morbidity.

The rabbit ear model simulates the human condition relatively well. Lymphoscintigraphy demonstrates the expected pattern of dermal backflow that

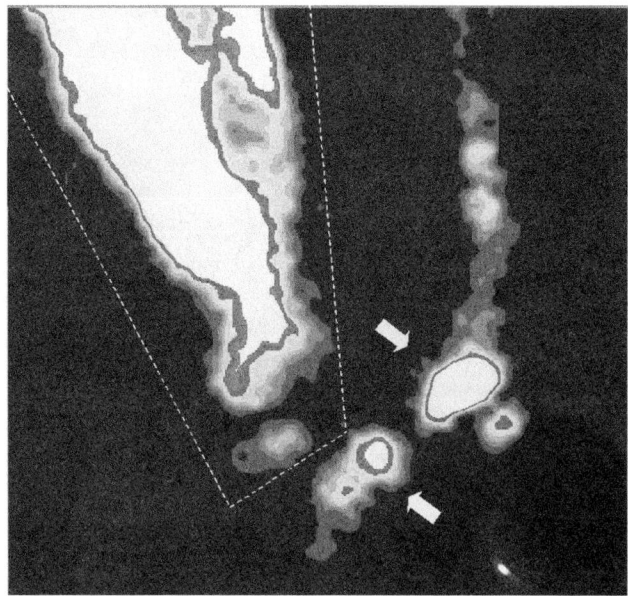

FIGURE 3. Radionuclide lymphoscintigraphy in the rabbit ear model of chronic lymphedema. The *dotted line* encloses the portion of the image that corresponds to the postsurgical, edematous ear. The *arrows* demarcate radiotracer in the proximal lymph nodes. In comparison to the normal ear, the postsurgical side shows a remarkable degree of dermal backflow, reduced lymphatic clearance, and absence of radionuclide delineation of lymph node drainage. These findings simulate the characteristics of lymphoscintigraphy in human acquired lymphedema.

accompanies diminished lymphatic clearance (FIG. 3). Furthermore, light microscopic assessment of postmortem specimens discloses the expected thickening of dermal and epidermal structures in the lymphedematous ear when compared to the normal ear (FIGS. 4A and B).

We have utilized this model of chronic postsurgical lymphedema to assess the therapeutic effect of exogenously administered recombinant human VEGF-C.[44] After complete surgical healing, a group of these animals received a single, subcutaneously administered dose of VEGF-C, with sacrifice and assessment one week following therapy. A group of control subjects received parenteral saline administration. With VEGF-C, there was statistically significant improvement in the lymphoscintigraphically assessed dynamic lymphatic function. Immunohistochemical analysis of the postmortem specimens disclosed evidence of a significant increase in lymphatic vascularity in the skin, coupled with a reversal of the hypercellularity that characterized the untreated lymphedematous state (FIG. 5).

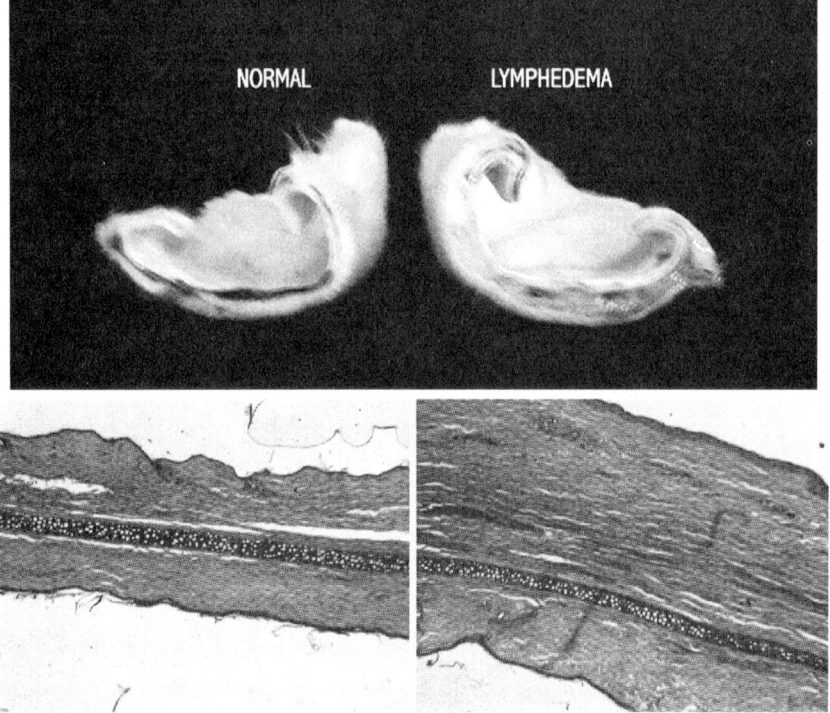

FIGURE 4. (**A**) Postmortem cross-sections of the ears from a representative rabbit show the expected thickening of the tissues in the lymphedematous ear. (**B**) On microscopic section, the lymphedematous ear (*right*) demonstrates hypercellularity and dermal and epidermal thickening when compared to the normal ear (*left*).

CONCLUSIONS

The central role of VEGF-C and the VEGFR-3 receptor in lymphatic development has become well established in recent years. The Chy mouse model demonstrates the utility of VEGF-C gene therapy in a model of human primary lymphedema. Preliminary data in a model of acquired lymphedema complement these observations and, hopefully, open the door to continued investigation application of such principles to the secondary forms of human lymphatic disease states.

Intensive future investigation will be required to establish dose–response relationships, along with the temporal nature and the durability of the therapeutic response. As with other forms angiogenic therapy, the relative virtues of growth factor therapy versus gene therapy must be established.

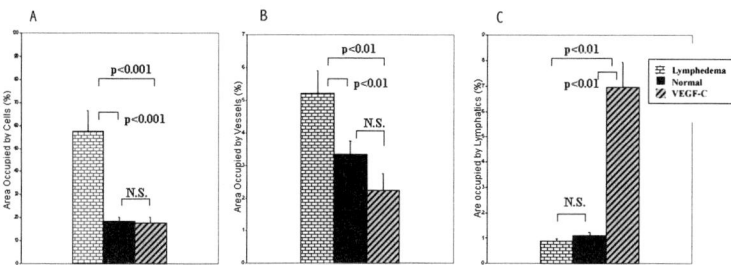

FIGURE 5. (A) The cutaneous thickening of the lymphedematous tissues is characterized by a dramatically increased cellular density in the dermis when compared to the skin of the normal ears (57.53 ± 8.82 vs. 18.16 ± 1.8; $P < 0.001$). These changes are consistent with the patterns observed in human chronic lymphedema. Treatment with VEGF-C restored the cellularity of the dermis to normal levels (17.51 ± 2.4 vs. 18.16 ± 1.8, respectively, NS) and resulted in normalization of the thickness of the epidermis. (B) Vascularity was analyzed through double-immunofluorescent staining. In lymphedema, the area occupied by blood vessels was generally increased (5.21 ± 0.7 vs. 3.34 ± 0.4, $P < 0.01$). Overall, blood vascularity of VEGF-C-treated specimens was not significantly different from that of normal skin (2.24 ± 0.51 vs. 3.34 ± 0.4, respectively, NS). (C) After VEGF-C treatment, there was, reproducibly, a significant increase in lymphatic vascularity, when compared to the saline-treated lymphedema specimens (6.97 ± 0.97 vs. 0.9 ± 0.08, $P < 0.01$; normal 1.1 ± 0.12, not significant when compared to lymphedema).

ACKNOWLEDGMENTS

The author is grateful to the Western States Affiliates of the American Heart Association for funding of the work cited.

REFERENCES

1. SZUBA, A. & S. ROCKSON. 1997. Lymphedema: anatomy, physiology and pathogenesis. Vasc. Med. **2:** 321–326.
2. ROCKSON, S. 2000. Primary lymphedema. *In* Current Therapy in Vascular Surgery. C. Ernst & J. Stanley, Eds. Mosby. Philadelphia, PA.
3. ROCKSON, S.G. 2001. Lymphedema. Am. J. Med. **110:** 288–295.
4. PASSIK, S., M. NEWMAN, M. BRENNAN, *et al.* 1995. Predictors of psychological distress, sexual dysfunction and physical functioning among women with upper extremity lymphedema related to breast cancer. Psycho-Oncology **4:** 255–263.
5. ROCKSON, S.G. 2002. Lymphedema after surgery for cancer: the role of patient support groups in patient therapy. Dis. Mangm. Health Outcomes 10: 345–347.
6. SZUBA, A. & S.G. ROCKSON. 1998. Lymphedema: classification, diagnosis and therapy. Vasc. Med. **3:** 145–156.

7. RIVARD, A. & J.M. ISNER. 1998. Angiogenesis and vasculogenesis in treatment of cardiovascular disease. Mol. Med. **4:** 429–440.
8. BAUMGARTNER, I., A. PIECZEK, O. MANOR, et al. 1998. Constitutive expression of phVEGF165 after intramuscular gene transfer promotes collateral vessel development in patients with critical limb ischemia. Circulation 97: 1114–1123.
9. FERRARA, N. & K. ALITALO. 1999. Clinical applications of angiogenic growth factors and their inhibitors. Nat. Med. **5:** 1359–1364.
10. PULLINGER, D. & H. FLOREY. 1937. Proliferation of lymphatics in inflammation. J. Pathol. Bacteriol. **45:** 157–170.
11. OLOFSSON, B., M. JELTSCH, U. ERIKSSON, et al. 1999. Current biology of VEGF-B and VEGF-C. Curr. Opin. Biotechnol. **10:** 528–535.
12. VEIKKOLA, T., M. KARKKAINEN, L. CLAESSON-WELSH, et al. 2000. Regulation of angiogenesis via vascular endothelial growth factor receptors. Cancer Res. **60:** 203–212.
13. JOUKOV, V., K. PAJUSOLA, A. KAIPAINEN, et al. 1996. A novel vascular endothelial growth factor, VEGF-C, is a ligand for the Flt4 (VEGFR-3) and KDR (VEGFR-2) receptor tyrosine kinases. EMBO J. **15:** 1751.
14. KAIPAINEN, A., J. KORHONEN, T. MUSTONEN, et al. 1995. Expression of the fms-like tyrosine kinase 4 gene becomes restricted to lymphatic endothelium during development. Proc. Natl. Acad. Sci. USA **92:** 3566–3570.
15. JELTSCH, M., A. KAIPAINEN, V. JOUKOV, et al. 1997. Hyperplasia of lymphatic vessels in VEGF-C transgenic mice. Science **276:** 1423–1425.
16. OH, S.J., M.M. JELTSCH, R. BIRKENHAGER, et al. 1997. VEGF and VEGF-C: specific induction of angiogenesis and lymphangiogenesis in the differentiated avian chorioallantoic membrane. Dev. Biol. **188:** 96–109.
17. SLEEMAN, J.P., J. KRISHNAN, V. KIRKIN, et al. 2001. Markers for the lymphatic endothelium: in search of the holy grail? Microsc. Res. Tech. **55:** 61–69.
18. ENHOLM, B., T. KARPANEN, M. JELTSCH, et al. 2001. Adenoviral expression of vascular endothelial growth factor-C induces lymphangiogenesis in the skin. Circ. Res. **88:** 623–629.
19. HALSTED, W.S. 1921. The swelling of the arm after operation for cancer of the breast, elephantiasis chirurgica, its causes and prevention. Bull. J. Hopk. Hosp. **32:** 309–313.
20. OLSZEWSKI, W., Z. MACHOWSKI, J. SOKOLOWSKI, et al. 1968. Experimental lymphedema in dogs. J. Cardiovasc. Surg. (Torino) **9:** 178–183.
21. SEGERSTROM, K., P. BJERLE, S. GRAFFMAN, et al. 1992. Factors that influence the incidence of brachial oedema after treatment of breast cancer. Scand. J. Plast. Reconstr. Surg. Hand. Surg. **26:** 223–227.
22. ERICKSON, V., M. PEARSON, P. GANZ, et al. Arm edema in breast cancer patients. 2001. J. Natl. Cancer Inst. **93:** 96–111.
23. WANG, G.Y. & S.Z. ZHONG. 1985. A model of experimental lymphedema in rats' limbs. Microsurgery **6:** 204–210.
24. LEE-DONALDSON, L., M.H. WITTE, M. BERNAS, et al. 1999. Refinement of a rodent model of peripheral lymphedema. Lymphology **32:** 111–117.
25. BOLLINGER, A. 1993. Microlymphatics of human skin. Int. J. Microcirc. Clin. Exp. **12:** 1–15.
26. LEU, A.J., D.A. BERK, F. YUAN, et al. 1994. Flow velocity in the superficial lymphatic network of the mouse tail. Am. J. Physiol. **267:** H1507–H1513.

27. SLAVIN, S.A., A.D. VAN DEN ABBEELE, A. LOSKEN, et al. 1999. Return of lymphatic function after flap transfer for acute lymphedema. Ann. Surg. **229:** 421–427.
28. SWARTZ, M.A., D.A. BERK & R.K JAIN. 1996. Transport in lymphatic capillaries. I. Macroscopic measurements using residence time distribution theory. Am. J. Physiol. **270:** H324–H329.
29. SWARTZ, M.A., A. KAIPAINEN, P.A. NETTI, et al. 1999. Mechanics of interstitial-lymphatic fluid transport: theoretical foundation and experimental validation. J. Biomech. **32:** 1297–1307.
30. BANERJI, S., J. NI, S.X. WANG, et al. 1999. LYVE-1, a new homologue of the CD44 glycoprotein, is a lymph-specific receptor for hyaluronan. J. Cell Biol. **144:** 789–801.
31. WIGLE, J.T. & G. OLIVER. 1999. Prox1 function is required for the development of the murine lymphatic system. Cell **98:** 769–778.
32. WIGLE, J.T., N. HARVEY, M. DETMAR, et al. 2002. An essential role for Prox1 in the induction of the lymphatic endothelial cell phenotype. EMBO J. **21:** 1505–1513.
33. WENINGER, W., T.A. PARTANEN, S. BREITENEDER-GELEFF, et al. 1999. Expression of vascular endothelial growth factor receptor-3 and podoplanin suggests a lymphatic endothelial cell origin of Kaposi's sarcoma tumor cells. Lab. Invest. **79:** 243–251.
34. PARTANEN, T.A. & K. PAAVONEN. 2001. Lymphatic versus blood vascular endothelial growth factors and receptors in humans. Microsc. Res. Tech. **55:** 108–121.
35. SCHOPPMANN, S.F., P. BIRNER, P. STUDER, et al. 2001. Lymphatic microvessel density and lymphovascular invasion assessed by anti-podoplanin immunostaining in human breast cancer. Anticancer Res. **21:** 2351–2355.
36. SZUBA, A. & S.G. ROCKSON. 1998. Lymphedema: A review of diagnostic techniques and therapeutic options. Vasc. Med. **3:** 145–156.
37. SZUBA, A., H. STRAUSS, S. SIRSIKAR, et al. 2002. Prognostic value of quantitative radionuclide lymphoscintigraphy in breast cancer-related lymphedema of the upper extremity. Nucl. Med. Commun. In press.
38. KARKKAINEN, M.J., R.E. FERRELL, E.C. LAWRENCE, et al. 2000. Missense mutations interfere with VEGFR-3 signalling in primary lymphoedema. Nat. Genet. **25:** 153–159.
39. KARKKAINEN, M.J. & T.V. PETROVA. 2000. Vascular endothelial growth factor receptors in the regulation of angiogenesis and lymphangiogenesis. Oncogene **19:** 5598–5605.
40. KARKKAINEN, M.J., L. JUSSILA, R.E. FERRELL, et al. 2001. Molecular regulation of lymphangiogenesis and targets for tissue oedema. Trends Mol. Med. **7:** 18–22.
41. MAKINEN, T., L. JUSSILA, T. VEIKKOLA, et al. 2001. Inhibition of lymphangiogenesis with resulting lymphedema in transgenic mice expressing soluble VEGF receptor-3. Nat. Med. **7:** 199–205.
42. KARKKAINEN, M.J., A. SAARISTO, L. JUSSILA, et al. 2001. A model for gene therapy of human hereditary lymphedema. Proc. Natl. Acad. Sci. USA **98:** 12677–12682.
43. HUANG, G.K. & Y.P. HSIN. 1983. An experimental model for lymphedema in rabbit ear. Microsurgery **4:** 236–242.
44. SZUBA, A., M. SKOBE, M.J. KARKKAINEN, et al. 2002.Therapeutic lymphangiogenesis with human recombinant VEGF-C. FASEB J. Oct. 18, 2002. 10,1096/fj02–0401fje.

Part 2: Biological Principles
Panel Discussion

The Effect of Experimental Lymphangiogenesis on Tissue Volume

QUESTION: Did you also assess whether this treatment reduces the 60 percent increase in tissue volume that you demonstrated in the absence of treatment?

ROCKSON: Indirectly, yes. The problem that we have wrestled with is that, quite frankly, volumetry by water displacement has been suitable to identify the rare animal subjects that do not develop measurable edema after surgery, but it doesn't have the ability to discriminate what are, honestly, the relatively modest volume changes in response to therapeutic lymphangiogenesis. I suspect that, in fact, we have been using a reasonably sub-therapeutic dose of VEGF-C within these studies.

Temporal Responsiveness to Exogenous VEGF-C

SUMNER SLAVIN (*Harvard Medical School*): In our own work with flaps in a rodent model, we have noticed that acute lymphedema is reversible by importing healthy lymphatics to the diseased area. Your development of a model using the VEGF-C is particularly intriguing. Would you tell us, based on your preliminary data, whether VEGF-C must be administered very early after the onset of the lymphedema, or whether you could perhaps treat a more chronically lymphedematous rabbit ear?

ROCKSON: That's an excellent question. We are very interested in the temporal nature of the lymphangiogenic response. In our so-called chronic lymphedema model, which reflects a 4-week duration of lymphedema, the appearance of the tissues at 4 weeks correlates very precisely to the appearance in rabbits that we have followed for as long as 6 to 12 months. Therefore, I interpret the model to reflect the equivalent of, for example, a patient who might develop overt postmastectomy lymphedema as early as 6 months or as late as 2 to 3 years after the surgical insult. So it is what I would conceive of as chronic lymphedema.

To the extent that there is not a substantial fibrotic response within this model, yet distinct architectural changes are observed, I would conjecture that we have therefore not identified a limit of responsiveness. Certainly, if this were to progress to a significant fibrotic response in the cutaneous tissues, the ability to respond may be substantially diminished, but with the cellular changes of the type that we have described, I think that it is plausible to hope that there might be a protracted responsiveness to VEGF-C.

The Mechanism of VEGF-C Responses in Experimental Lymphedema

MICHAEL DETMER (*Harvard University*): Your model is extremely interesting. In your opinion, how does the VEGF-C work in this model to reduce the lymphedema? As I understand it, you excise the strip of skin and the lymphatics, so basically you have no bridging lymphatics between the tip of the ear and the rest of the body. So, does VEGF-C treatment then induce new growth of lymphatics that would bridge over the previous wound area, and does that not occur in the normal situation?

ROCKSON: I can certainly advance my hypothesis. One of the reasons that I think that this model is, in fact, a very useful tool for the sort of questions that we are asking is that the utility of the model may relate to interspecies variation in the vigor of the lymphangiogenic response to wound healing. We know, for example, that for the smaller rodents, a very substantial surgical insult is required to produce lymphedema because endogenous lymphatic regeneration occurs quite readily and successfully. In contrast, I think that the rabbit may have an inherently less successful, less vigorous endogenous lymphangiogenic response. And, because the rabbit ear is not a region that is richly endowed with lymphatics at the outset, this may render the model even more useful.

The Effect of VEGF-C on Fibrosis and Adipose Tissues in Lymphedema

ALBERT MILLER (*Northwestern University*): Would you speculate on how VEGF-C might function in the patient who has had longstanding lymphedema, where there is a great deal of fat under the skin, as well as the fibrosis that you mentioned.

ROCKSON: I believe that it has been established that there is an inverse relationship between the rate of lymphatic transport out of a tissue region and the amount of adipose accumulation in the tissue. Speculatively, one can surmise that this is related to the process of adipogenesis, but it may simply reflect hypertrophy of preexisting adipose cells.

Therefore, if, late in the course of the disease, after adipose accumulation occurs, one were able to reestablish lymphatic drainage, successful therapeutic lymphangiogenesis should, in principle, have the ability to reverse the adiposity if the relationship between lymphatic flow and adipose tissue stimulation is dynamic in both directions. Deriving an answer to this question is very important, but will demand a lot of intense investigation and further delineation of molecular mechanisms, as well.

Gene Mutations Observed in Primary Lymphedema Families

CARLA MOUTA (*Maine Medical Center*): Dr. Ferrell, in your summary, you suggested that you had eliminated an association between a series of genes like LYVE-1, PROX-1, and others, with lymphedema. I didn't understand whether you implied that there were mutations within those genes that had nothing to do with lymphedema, or whether you were looking at the expression level of those genes in patients that have lymphedema mutations.

FERRELL: I did not intend to leave the impression that we have eliminated those genes as a cause of lymphedema. We have not found mutations in any of those genes that co-segregate with lymphedema in the affected families. In virtually all of those genes, we have identified variation, often missense mutations in the genes themselves, but these are found at a comparable frequency in the general population. Therefore, the variation does not seem to be responsible for the lymphedema phenotype. It would be incorrect to say that we have eliminated those genes, because there is always the possibility that there are mutations within the gene or regulatory mutations that we have not detected. At this stage, we have no gene expression data from patients.

MOUTA: Have you looked to see whether FOXC2 regulates the expression of any of those genes? I'm primarily interested in PROX-1, since both are transcription factors.

FERRELL: We have not.

The Aging Effect in Lymphatics

MOUTA: Dr. Zawieja, it seems that a lot of these diseases are age-related. In my investigation of the effects of aging on lymphatics, I encountered the work that you did in collaboration with the University of St. Petersburg in Russia. Was there some mention that, at least in the mesentery, the number of lymphatics decreases over time?

ZAWIEJA (*Texas A&M University*): There is, within the older Russian literature, an unequivocal description of the aging of lymphatics in the heart. With dye injection at the time of coronary bypass surgery, the lymphatics diminish in number and become more tortuous with age.

Clinical Trials of VEGF-C Therapy

SIMON SIMONIAN (*Georgetown University*): Dr. Rockson, when will VEGF-C come to the bedside from the bench?

ROCKSON: I think that the pressing question is, what the impact will be, in the clinical setting, when there is a history of prior malignancy. If it is our intent to treat secondary forms of lymphedema, we would be segregating a patient population that by its nature is made up of malignancy survivors. I have the inherent belief that, provided that one screens properly for the presence of active malignant disease, there will still be an acceptable margin of safety, but it is a question that will need to be addressed, obviously, and certainly one that would begin to be addressed in Phase I testing.

Upregulation of VEGF-C in Lymphedema?

GAVIN THURSTON (*Regeneron Pharmaceuticals*): When tissue is hypoxic, there is an upregulation of VEGF-A and the VEGFR2 receptor, so I wonder whether there a corresponding upregulation of VEGF-C in tissue? Does tissue "sense" that it needs more lymphatic growth? And, if so, have you looked at VEGF-C expression in the rabbit to see if it is upregulated? And, if it is upregulated, or if there is any change in VEGF-C, do you think you can affect the VEGF-C response by giving more VEGF-C?

ROCKSON: I don't have data to share, but I can certainly hypothesize. I harbor the conviction, at least in the human disease, that some unidentified biological attribute distinguishes the subset of individuals that are predisposed to develop overt lymphedema after a traumatic event such as axillary lymph node dissection or irradiation. In the case of postmastectomy lymphedema, it is estimated that perhaps only 20–25% of the women exposed to a comparable surgical insult will develop clinically significant lymphatic dysfunction.

I believe that there is a variability, presumably genetically endowed, to the vigor of the lymphangiogenic response, which, in turn, might translate into the vigor with which there is upregulation, for example, of the VEGFR3 receptor. This phenomenon has not yet been adequately investigated. With regard to our specific mode, we have not yet examined expression patterns of VEGF-C in the tissues. Of course, it is something that we are interested in exploring.

Frequency of Human VEGFR-3 Mutation

THOMAS GLOVER (*University of Michigan*): Dr. Ferrell, in your studies and others, what percent of Milroy's patients have mutations of the VEGFR3?

FERRELL: We haven't actually segregated our families into those with a congenital onset in order to estimate the proportion in which we found VEGFR3 mutations. Overall, it involves about six percent of all of the primary lymphedema families that we're studying.

Placental Growth Factor (PlGF) and Its Receptor Flt-1 (VEGFR-1)

Novel Therapeutic Targets for Angiogenic Disorders

AERNOUT LUTTUN, MARC TJWA, AND PETER CARMELIET

Center for Transgene Technology and Gene Therapy, Flanders Interuniversitary Institute for Biotechnology, Leuven, Belgium

ABSTRACT: Efforts to therapeutically stimulate or inhibit vessel growth have been primarily focused on vascular endothelial growth factor (VEGF) and its receptor VEGFR-2 (Flk-1), while little attention has been devoted to the therapeutic potential for angiogenic disorders of placental growth factor (PlGF), a VEGF family member, and its receptor VEGFR-1 (Flt-1). However, recent developments and insights could shift that focus to P1GF and Flt-1. Indeed, PlGF stimulated angiogenesis and collateral growth in ischemic heart and limb with at least a comparable efficiency to VEGF and did not cause side effects associated with VEGF, such as edema or hypotension. An anti-Flt-1 antibody suppressed neovascularization in tumors and ischemic retina, and angiogenesis and inflammatory joint destruction in arthritis. The anti-Flt-1 antibody also reduced atherosclerotic plaque growth and vulnerability, but the atheroprotective effect was not due to reduced plaque neovascularization. The anti-inflammatory effects of the anti-Flt-1 antibody were attributable to a reduced mobilization of bone marrow–derived myeloid progenitors into the peripheral blood, a reduced mobilization/differentiation (and impaired infiltration) of Flt-1-expressing leukocytes into inflamed tissues, and a defective activation of myeloid cells. Thus, PlGF and Flt-1 constitute potential candidates for therapeutic modulation of angiogenesis and inflammation.

KEYWORDS: PlGF; Flt-1; angiogenesis; inflammation; collateral growth; ischemia; retinopathy; cancer; atherosclerosis; rheumatoid arthritis

---

Address for correspondence: P. Carmeliet, M.D., Ph.D., Center for Transgene Technology and Gene Therapy, Flanders Interuniversitary Institute for Biotechnology, KULeuven, Campus Gasthuisberg, Herestraat 49, B-3000, Leuven, Belgium. Voice: 32-16-34.57.72; fax: 32-16-34.59.90.
peter.carmeliet@med.kuleuven.ac.be

Ann. N.Y. Acad. Sci. 979: 80–93 (2002). © 2002 New York Academy of Sciences.

INTRODUCTION

Vascular endothelial growth factor (VEGF) and its receptor tyrosine kinase-2 (VEGFR-2 or Flk-1/KDR) have received wide attention for therapeutic stimulation or inhibition of angiogenesis.[1] VEGF, however, induces only transient and modest improvement of cardiac function in some (but not all) patients with ischemic heart disease, suggesting that additional angiogenic agents may be required.[2,3] Inhibitors of VEGF and its receptors are currently under evaluation for treatment of cancer and other disorders characterized by excess blood vessel growth.[4] Although VEGF binds both receptor tyrosine kinases Flk-1 and Flt-1 (VEGFR-1), inhibitors have been primarily targeted to neutralization of Flk-1 activity, based on the assumption that VEGF-driven angiogenesis is primarily mediated via Flk-1.[1,5]

The role of Flt-1, to which not only VEGF binds, but also its homologues PlGF and VEGF-B, has remained enigmatic, primarily because Flt-1 has low tyrosine kinase activity[6,7] and its signaling pathways remain poorly characterized.[8] Mouse embryos lacking Flt-1 succumb because of vascular defects, while mice expressing Flt-1 lacking the tyrosine kinase domain survive,[6,9] suggesting that, during development, Flt-1 primarily functions as a non-signaling "reservoir" for VEGF. By displacing VEGF from Flt-1, PlGF provides additional VEGF to induce Flk-1-mediated angiogenic signaling.[10] Recent gene-targeting studies indicate, however, that Flt-1-mediated signaling may play a significant role in pathological angiogenesis. Indeed, loss of PlGF impairs angiogenesis in the ischemic retina, limb, and heart, in wounded skin, and in cancer, without affecting physiological angiogenesis.[11] Furthermore, genetic truncation of the Flt-1 tyrosine kinase domains[12] or antisense-mediated downregulation of Flt-1 suppresses tumor angiogenesis and VEGF-induced angiogenesis.[13] Since PlGF and Flt-1 have a restricted angiogenic activity in pathological conditions, they may be attractive therapeutic candidates for pro- or anti-angiogenesis. Therefore, the therapeutic potential of PlGF for stimulation of angiogenesis in ischemic tissues and of Flt-1-inhibitors for nhibition of prototypic angiogenic disorders such as cancer, retinal ischemia, arthritis, and atherosclerosis was studied in the present report.

PlGF STIMULATES REVASCULARIZATION OF THE ISCHEMIC MYOCARDIUM AND LIMB

Myocardial ischemia results from an imbalance between the supply and consumption of oxygen. Initially, the myocardium develops a protective response by adapting a state of hibernation to preserve high-energy metabolites at the expense of contractile dysfunction. At this stage, the hibernating myocardium is still viable and able to restore its contractile function upon proper revascularization. However, when the ischemic insult becomes too severe or lasts too long, the hibernating myocardium may undergo irreversible structural changes and degenerate, to become replaced by fibrotic scar tissue. Genetic studies performed in our laboratory in transgenic mice lacking two isoforms of VEGF indicated that impaired myocardial angiogenesis led to ischemic heart disease with signs of hibernation that, over time, progressed to cardiac failure.[14] The formation not only of endothelial-lined capillaries ("angiogenesis") but also of coronary arteries ("arteriogenesis") was impaired. These studies provided genetic evidence that insufficient availability of an angiogenic growth factor results in ischemic heart disease. They also constituted a rationale to prevent or rescue hibernating myocardium by stimulating myocardial angiogenesis via administration of angiogenic growth factors. Preclinical animal studies have indeed documented favorable effects of stimulating growth of myocardial capillaries or expansion of pre-existing collaterals after administration of VEGF or basic fibroblast growth factor (bFGF), either as recombinant protein or by gene transfer.[2,3] Initial clinical experience with therapeutic angiogenesis and arteriogenesis in ischemic heart patients also raises hope that such strategies may be useful to improve cardiac function.[15]

For therapeutic revascularization of ischemic tissues to be effective, new vessels should be mature and durable, so that they do not regress once delivery of the therapeutic angiogenic factor is halted. PlGF-induced vessel growth was studied by adenoviral PlGF gene transfer into the skin of ears.[16] Control vectors failed to affect skin vasculature, but PlGF gene transfer caused pre-existing vessels to enlarge, resulting in tortuous, thin-walled, pericyte-poor "mother" vessels. The latter vessels remained enlarged and subsequently stabilized into mature, durable vessels by acquisition of a pericyte coat and, occasionally, by deposition of a thin rim of perivascular collagen. Remarkably, these vessels remained functional and persisted for more than one year after PlGF gene transfer, even though transgenes are only transiently expressed for 4 weeks in this model.[16] While both PlGF and VEGF stimulated the formation of large stabilized vessels, PlGF avoided the harmful complications of VEGF such as edema, fibrin deposition, and growth of unstable vascular tangles and glomeruloid bodies ("hemangiomagenesis") (FIG. 1).[16,17]

The therapeutic potential of PlGF to stimulate angiogenesis was evaluated in a mouse model of ischemic myocardial revascularization. Since both VEGF and PlGF exist as multiple isoforms that differ in molecular mass, solubility, and receptor binding, the effect of the PlGF isoforms hPlGF-1, hPlGF-2, and mPlGF-2 was compared to that of the VEGF isoforms $hVEGF_{165}$ and $hVEGF_{121}$. All forms of VEGF and PlGF stimulated the growth of new vessels ("angiogenesis"). PlGF may induce myocardial revascularization by amplifying the angiogenic activity of VEGF, which is upreg-

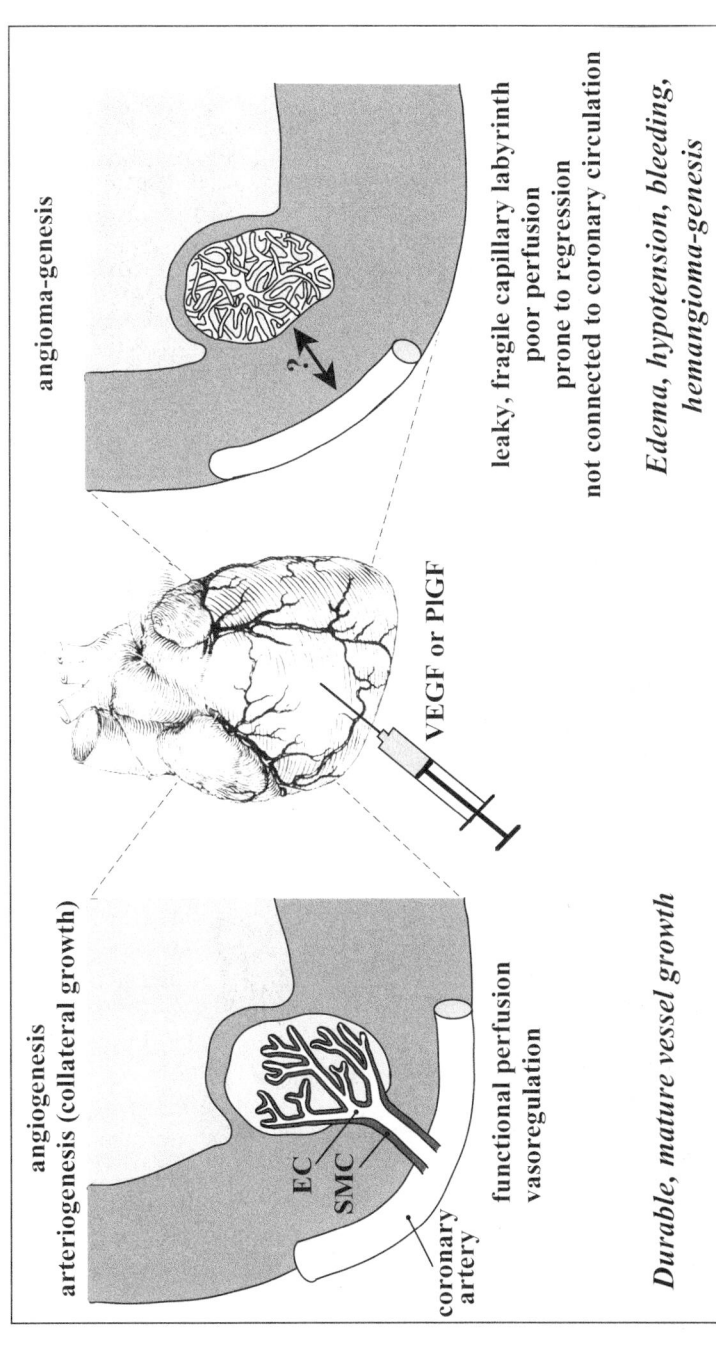

FIGURE 1. Therapeutic angiogenesis and arteriogenesis with PlGF in the ischemic myocardium and limb. VEGF and PlGF efficiently stimulate angiogenesis but, compared to PlGF, VEGF induces more hypotension, edema, leakiness, and "hemangioma-genesis." In addition, PlGF-induced vessels are durable and mature, and persist for prolonged periods after arrest of PlGF delivery.

ulated in the ischemic myocardium.[11] In addition, PlGF stimulated myocardial angiogenesis by increasing VEGF expression by fibroblasts, which are abundant in the myocardial stroma. PlGF also stimulated "arteriogenesis," or the maturation of vessels via coverage with smooth muscle cells (SMCs) leading to stabilization and durability of new vessels (FIG. 1), as 25 to 30% of the new myocardial vessels stained positively for the SMC α-actin marker after treatment with all PlGF isoforms. Since a similar fraction of myocardial vessels is normally covered by SMCs (25%), PlGF did not cause "hemangioma-genesis" but created a new myocardial vasculature with normal characteristics (FIG. 1). The functionality of the new vasculature was evidenced by the improved perfusion of the ischemic myocardium after PlGF treatment. While PlGF and VEGF might affect SMCs indirectly via release of SMC mitogens from activated endothelial cells (ECs), they could also stimulate SMCs directly, since both Flt-1 and Flk-1 are present on SMCs. PlGF determined the responsiveness of SMCs to VEGF, since PlGF-deficient SMCs proliferated only normally in response to VEGF when PlGF was present. Thus, PlGF stimulated myocardial revascularization as efficiently as VEGF but, unlike VEGF, did not cause edema or hypotension (FIG. 1).

The potential of PlGF to stimulate the growth of pre-existing arterial collaterals and their second- and third-generation side branches ("collateral growth") was evaluated by treating mice with PlGF after ligation of their femoral artery. Delivery of 1.5 µg hPlGF-2 per day minimally affected the primary collaterals and capillaries in the adductor region, but significantly increased the number and size of the second- and third-generation collateral branches, thereby enlarging the collateral perfusion area (sum of the luminal areas of all secondary and tertiary collaterals). Also, PlGF treatment increased the angiographic number of precapillary arterioles that regulate vascular resistance and tissue perfusion (for example, the collateral side branches >300 µm^2). As a result, hindlimb perfusion, determined by laser doppler or microspheres, was increased more than 3-fold by PlGF treatment. Importantly, PlGF treatment improved the spontaneous mobility and the functional muscle reserve of the ischemic hind limb, evaluated using a novel swim endurance exercise test. In contrast, a similar dose of hVEGF$_{165}$ was less efficient and consistent in inducing structural and perfusional changes and did not improve the functional swim endurance reserve. The arteriogenic mechanisms of PlGF appeared to be mediated by effects on SMCs and macrophages, known to produce SMC/EC-mitogens and cytokines during collateral growth.[18,19] Indeed, Flt-1 expression was upregulated in collateral vessels after femoral artery ligation, while PlGF treatment stimulated SMC growth and macrophage infiltration, as evidenced by the increased muscular thickness of the collateral branches and recruited macrophages around the collateral side branches. Moreover, PlGF upregulated the production by macrophages of TNFα and MCP-1, cytokines implicated in collateral growth.[20,21] Thus, compared to VEGF, PlGF more efficiently stimulated the functional recovery of the ischemic limb, primarily by enhancing growth of collateral side branches.

Taken together, these recent data uncover some therapeutically attractive mechanisms of PlGF for post-ischemic revascularization. Flt-1 and its ligands may stimulate EC growth via several complementary and mutually nonexclusive mechanisms (see subsequent text). PlGF and Flt-1 also stimulate "arteriogenesis" by affecting smooth muscle cells (SMCs), which are critical for the establishment of mature, durable, and functional neovessels.[18,19] Indeed, vessels covered with SMCs fail to regress and secure persistent tissue perfusion when the angiogenic stimulus fades away.[18] PlGF treatment may achieve this therapeutic goal as it combines two desirable properties, for example, it stimulates SMCs coincidently with ECs and thereby may induce vessel growth in a more balanced manner than VEGF or PDGF-BB, which preferentially stimulate ECs or SMCs, respectively.[22] In the ischemic myocardium, PlGF could recruit SMCs by itself[23] or, as illustrated by the present *in vitro* findings, in synergy with VEGF. The latter amplification of VEGF by PlGF might explain why PlGF stimulated revascularization only in ischemic but not in healthy tissues. Improved perfusion of ischemic tissues does not, however, rely only on the growth of small capillaries ("angiogenesis") and the maturation of these naked endothelial channels with an SMC-coat ("arteriogenesis"), but also on the growth of large arterial collaterals ("collateral growth"), since the latter conduct larger amounts of blood flow.[18,19] Collateral growth in the ischemic limb is dependent on the recruitment of inflammatory cells, which produce EC/SMC mitogens.[18,19] To date, only a few arteriogenic molecules have been identified to affect growth. Compared to bFGF[24] or MCP-1,[18,20,21] which preferentially affect either ECs/SMCs or inflammatory cells, respectively, PlGF affected the three principal cell types responsible for collateral growth[19]: ECs,[11] SMCs, and macrophages, which all expressed increased levels of Flt-1 during collateral growth. Notably, PlGF recruited and activated monocytes to produce increased amounts of MCP-1 and TNFα, cytokines implicated in collateral growth.[20,21] It remains to be determined whether PlGF also enhanced recruitment of myeloid and/or other progenitors; however, since transplantation of wild-type bone marrow rescued the impaired collateral growth in PlGF-deficient mice,[11] such a mechanism appears plausible.

Flt-1 ANTAGONISTS FOR INHIBITION OF ANGIOGENIC DISORDERS

Significant efforts are under way to develop angiogenesis inhibitors for the treatment of angiogenic disorders such as cancer, retinopathy, arthritis, and other inflammatory disorders. VEGF and its receptor Flk-1 have been considered primary targets, while Flt-1 has been largely neglected. To study whether inhibition of Flt-1 would inhibit angiogenesis in ischemic, malignant, and inflammatory disorders, a monoclonal anti-Flt-1 antibody MF1 (anti-Flt-1 mAb), which blocked binding of VEGF and PlGF to Flt-1 and VEGF- or PlGF-driven growth of ECs,[11] was compared to control IgG and to monoclonal anti-Flk-1 antibody DC101 (anti-Flk-1 mAb). Anti-Flt-1 mAb efficiently suppressed VEGF-driven neovascularization in the cornea and matrigel implants and blocked neovascularization in the ischemic retina to a degree comparable to genetic deficiency of PlGF[11] or inhibition of Flk-1.[25] Anti-Flt-1 mAb dose-dependently blocked angiogenesis and growth of human epidermoid A431 tumors in nude mice, and was only slightly less active than anti-Flk-1 mAb. Compared to the large, vascularized control tumors, anti-Flt-1 mAb-treated tumors were pale and poorly vascularized, and exhibited extensive necrosis, reduced proliferation, and increased tumor cell apoptosis. Flt-1 was expressed in tumor-associated vessels, but not in malignant tumor cells. After 2 weeks of treatment with anti-Flt-1 mAb, the microvessel density was reduced by 45% and the vessel size by 30%. Anti-Flt-1 mAb also attenuated the growth and vascularization of PlGF- or VEGF-transduced rat C6 gliomas implanted in nude mice. These data therefore indicate an important role of Flt-1 in pathological angiogenesis (FIG. 2).

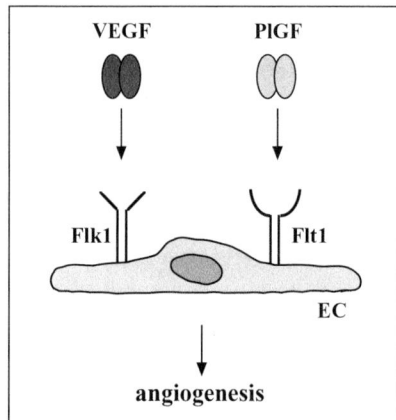 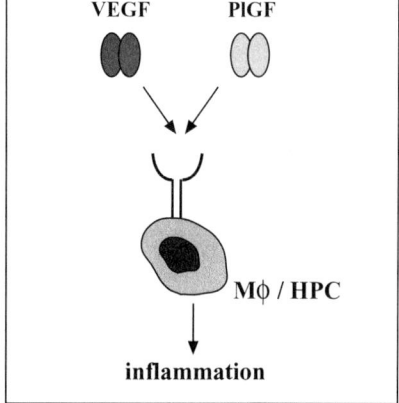

FIGURE 2. Angiogenesis driven by VEGF/PlGF can be mediated not only through Flk-1 but also through Flt-1, both present on endothelial cells (EC). However, unlike Flk-1, Flt-1 is also present on inflammatory cells (that is, macrophages, indicated by mf within the figure) and their progenitors (hematopoietic progenitor cells, indicated by HPC). Therefore, PlGF and VEGF can stimulate inflammation in addition to angiogenesis through Flt-1.

Flt-1 and its ligands may stimulate EC growth via several complementary and mutually non-exclusive mechanisms. First, by displacing VEGF from Flt-1, PlGF would make more VEGF available to activate Flk-1.[10] Second, PlGF could upregulate expression of VEGF by peri-endothelial fibroblasts, SMCs, or inflammatory cells in wound or tumor stroma (Bottemley et al.[26] and present findings). Third, Flt-1 could transmit its own intracellular angiogenic signals.[7,8] Fourth, PlGF might activate receptor cross-talk between Flt-1 and Flk-1, leading to enhanced Flk-1-driven angiogenesis (unpublished results). The regulation of EC functions by Flt-1 may also explain why inhibition of Flt-1 efficiently blocked VEGF-driven angiogenesis in the cornea, ischemic retina, arthritic joints, and tumors. The observation that anti-Flt-1 mAb suppressed pathological angiogenesis comparably to anti-Flk-1 mAb indicates that Flt-1 is a more important therapeutic target for inhibition of angiogenesis than previously presumed (FIG. 2). Others reported that anti-Flt-1 antibody did not affect tumor growth,[27] but insufficient amounts of antibody may have been used, since a more complete inhibition of Flt-1 reduced tumor angiogenesis in the present study. In addition, our data extend previous findings that anti-sense-mediated downregulation of Flt-1 suppressed VEGF-driven tumor angiogenesis.[13] Growth of Lewis lung carcinomas (LLCs), endogenously expressing VEGF or overexpressing PlGF, was also impaired in mice expressing Flt-1 without the tyrosine kinase domain (Flt-1/TK$^-$).[12] When VEGF was overexpressed in LLCs, tumor growth was restored, presumably because VEGF levels were sufficiently elevated to drive angiogenesis exclusively via Flk-1.[12] Thus, the relative importance of Flk-1- versus Flt-1-driven angiogenesis may depend on the relative expression of VEGF, PlGF, Flk-1, and Flt-1.

Flt-1 ANTAGONISTS FOR INHIBITION OF ANGIOGENIC AND INFLAMMATORY DISORDERS

Neovascularization of atherosclerotic plaques has been proposed to accelerate lesion growth and to make plaques more vulnerable to rupture, which could trigger fatal thrombotic complications (FIG. 3).[28,29] Angiogenesis inhibitors might be useful, but studies have been limited,[28] and the effect of VEGF receptor inhibitors has not been studied. We therefore studied initial (avascular) fatty streak lesions and advanced (vascular) complex plaques in atherosclerosis-prone apolipoprotein-E-deficient (ApoE$^{-/-}$) mice. ApoE$^{-/-}$ mice were treated with control IgG, anti-Flt-1, or anti-Flk-1 mAb for 5 weeks starting at 5, 10, or 20 weeks of age. Treatment with anti-Flt-1 mAb reduced the size of early and intermediate lesions at the aortic root by 50% and the growth of advanced atherosclerotic lesions by ~25%. Remarkably, the anti-Flk-1 mAb failed to affect atherosclerotic plaque development at all stages.

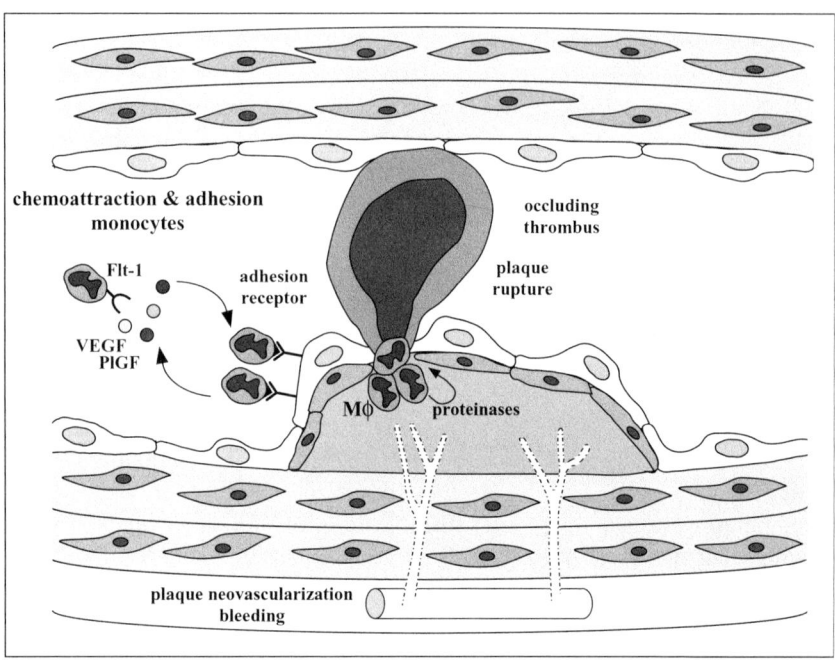

FIGURE 3. Role of VEGF and PlGF in atherosclerosis. VEGF or PlGF could promote atherosclerotic lesion growth and destabilization by stimulating: (i) recruitment and adhesion of monocytes; (ii) the production of proteolytic factors, thereby inducing degradation of the fibrous cap leading to plaque rupture; and (iii) plaque neovascularization, thereby facilitating monocyte infiltration. Since anti-Flt-1 mAb blocked atherosclerotic lesion growth independent of its anti-angiogenic effects, and anti-angiogenic anti-Flk-1 mAb were ineffective, it can be hypothesized that VEGF and PlGF promote atherosclerosis primarily through their effects on monocyte infiltration, rather than their effect on plaque neovascularization.

The atheroprotective effect of anti-Flt-1 mAb was attributable to a reduced macrophage infiltration in early as well as in advanced lesions (FIG. 3). Based on the anti-angiogenic effect of the anti-Flt-1 antibody in tumors and ischemic retina (see the preceding text), we had anticipated that the reduced plaque growth might, at least in part, be attributable to inhibition of plaque neovascularization (FIG. 3). Surprisingly, inhibition of neither Flt-1 nor Flk-1 blocked angiogenesis in atherosclerotic lesions or in the adventitia. VEGF and PlGF were expressed in plaques, but it remains to be determined whether their pro-angiogenic activity was overshadowed by other stimulators (bFGF, integrins) or counterbalanced by angiogenesis inhibitors (thrombospondin-1, tissue inhibitors of metalloproteinases [TIMPs]) in plaques.[28,30] Thus, anti-Flt-1 mAb suppressed plaque growth and vulnerability via inhibition of inflammatory cell infiltration, independently of angiogenesis, while anti-angiogenic anti-Flk-1 mAb, which normally blocks angiogenesis, was ineffective.

Angiogenesis may also contribute to the proliferation of synoviocytes, infiltration of inflammatory cells, cartilage destruction, and pannus formation—all hallmarks of rheumatoid arthritis.[31,32] VEGF is upregulated in arthritic joints and neutralization of VEGF reduces joint destruction,[33,34] but inhibitors of VEGF receptors have not been evaluated. To study the therapeutic potential of anti-Flt-1 and anti-Flk-1 mAb, polyarticular arthritis was induced in mice, using an autoimmune model of collagen type II-induced arthritis (a model for rheumatoid arthritis in humans[35]). Immunostaining of affected joints revealed that VEGF was present in inflammatory cells, chondrocytes. and cells at the pannus/bone interface and on ECs in synovial neovessels. Flk-1 was present only on synovial neovessels. Flt-1 and PlGF were expressed by inflammatory and, probably, also by ECs in the inflamed synovium. Treatment with anti-Flt-1 mAb reduced the incidence of joint disease by 60%, while all IgG-treated mice developed signs of arthritis in the paws and ankles. Remarkably, anti-Flt-1 mAb treatment suppressed the development of clinical symptoms (paw swelling, erythema, and ankylosis) by 85%. The effect observed with treatment by anti-Flt-1 mAb was specific, since anti-Flk-1 mAb was ineffective. Synovial infiltration by inflammatory cells was reduced by anti-Flt-1 mAb treatment. In addition, anti-Flt-1 mAb inhibited synovial angiogenesis. The anti-Flt-1 antibody suppressed activation of leukocytes and their production of TNFα and MCP-1, cytokines implicated in arthritis.[31] Thus, inhibition of Flt-1, but not of Flk-1, protected against arthritic joint destruction by suppressing synovial inflammation and neovascularization.

The failure of anti-Flk-1 mAb, but not of anti-Flt-1 mAb, to block arthritis and atherosclerosis and the angiogenesis-independent atheroprotective effect of the anti-Flt-1 mAb indicated that suppression of inflammation—not angiogenesis—was primarily responsible for the observed effects. Transplantation of bone marrow, transduced with a retroviral GFP-expressing vector, revealed that anti-Flt-1 mAb blocked the accumulation of GFP-labeled bone marrow-derived cells in atherosclerotic lesions. Reduced leukocyte accumulation could result from an effect of anti-Flt-1 mAb on the infiltration of circulating myeloid cells in inflamed lesions and/or from an effect on the differentiation or mobilization of these cells or their progenitors from the bone marrow into the peripheral blood (FIGS. 2 and 4). In support of the latter mechanism, anti-Flt-1 mAb partially abrogated the disease-associated increase in circulating monocytes and granulocytes. In addition, mobilization of hematopoietic progenitors into the peripheral blood was suppressed by the anti-Flt mAb by 75% (FIG. 4).

Flt-1, unlike Flk-1, is also expressed by inflammatory cells (this study and Sawano et al.[36]). The activity of PlGF and VEGF to attract Flt-1$^+$ leukocytes may explain why anti-Flt-1 antibodies suppressed inflammatory disorders,

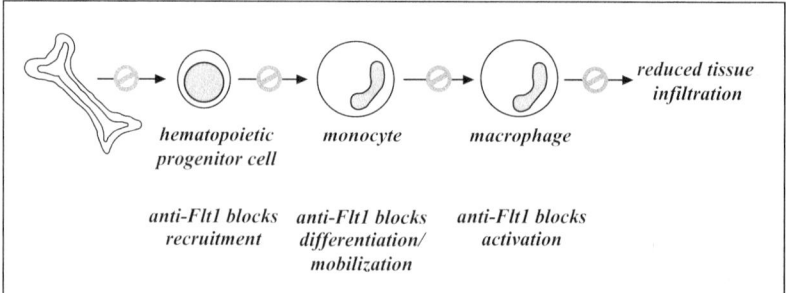

FIGURE 4. Flt-1 antagonists suppress inflammation by inhibiting the mobilization of myeloid progenitors from the bone marrow, impairing myeloid cell differentiation/mobilization, reducing their activation, and reducing cytokine production. As a result, blocking of Flt-1 leads to a reduced presence of activated macrophages in the inflamed target tissue, such as the atherosclerotic plaque or the rheumatoid joint.

while anti-Flk-1 antibodies were ineffective. PlGF and VEGF, which are upregulated within these inflamed tissues, may provide recruitment signals for differentiated inflammatory cells to home to and infiltrate in sites of ongoing inflammation (FIG. 2). However, they may also induce mobilization of myeloid progenitors from the bone marrow to the blood (FIGS. 2 and 4). Our findings that anti-Flt-1 antibody attenuated the disease-induced upregulation of circulating leukocytes and the granulocyte colony-stimulating factor (GC-SF)-induced mobilization of myeloid progenitors support such a mechanism and extend previous findings that anti-Flt-1 antibody blocked the recruitment of perivascular Flt-1$^+$ myeloid cells in tumors.[27] Flt-1 also appears to be involved in cellular activation (FIG. 4), since anti-Flt-1 antibody impaired the production of MCP-1 and TNFα by macrophages. An Flt-1-mediated increase in release of proteinases[37] might further explain why anti-Flt-1 mAb suppressed cartilage destruction in arthritic joints. Furthermore, since PlGF upregulates VEGF production by monocytes,[26] inhibition of Flt-1 would reduce VEGF-driven angiogenesis and, secondarily, inflammation. Taken together, the role of Flt-1 in mobilization of myeloid progenitors and in migration, activation, and (possibly) differentiation of myeloid cells may explain part of the therapeutic effect of anti-Flt-1 antibody in inflammatory disorders (FIG. 4).

MEDICAL IMPLICATIONS

Our findings may have several medical implications. First, inhibition of VEGF-driven angiogenesis has been considered to be an attractive therapy for inflammatory disorders.[4,28,38,39] However, inhibition of the prototype angiogenic Flk-1 receptor, which efficiently inhibits tumor angiogenesis, did not block arthritis or atherosclerosis, while Flt-1-inhibitors did. This implies that blocking Flk-1-driven angiogenesis alone, in the absence of anti-inflammatory agents, may not suffice. Second, anti-Flt-1 antibody did not only reduce the size but, more importantly, also stabilized atherosclerotic lesions, which might prevent plaque rupture and fatal thrombotic events.[40] This degree of protection by anti-Flt-1 mAb is comparable to—or even better than—that achieved by treatment with statins[40] or other anti-inflammatory compounds.[41,42] Since anti-Flt-1 antibody was only administered for 5 weeks, an increased efficiency might be possible with chronic administration. Third, inhibition of Flt-1 may provide a novel treatment for the inhibition of tumor angiogenesis and ischemic retinal neovascularization, and of inflammatory diseases. Lastly, PlGF may be an attractive candidate for improved therapeutic angiogenesis and arteriogenesis, as it efficiently stimulated revascularization in ischemic tissues without affecting quiescent vessels. Importantly, the PlGF-induced mature vessels persisted for prolonged periods (>1 year), even long after the arteriogenic stimulus had disappeared, indicating that it may suffice to stimulate new vessels with a short-term delivery of PlGF. Moreover, PlGF is less likely to cause the side-effects in situations where chronic, sustained delivery may be required,[15,43] since it did not cause undesirable side-effects associated with VEGF, like hyperpermeability, edema, and hemangioma-genesis (FIG. 1). Obviously, as for all pro- or anti-angiogenic compounds, caution is warranted to avoid aggravation of ischemic tissue disease by uncontrolled use of anti-Flt-1 antibody, or of atherosclerosis or cancer by systemic delivery of PlGF. While future studies will need to address in more detail the safety versus efficacy of Flt-1-ligands and its inhibitors in other preclinical models, anti-Flt-1 mAb did not appear to impair myocardial angiogenesis, wound healing, or ovulation (results not shown). In conclusion, these findings provide a rationale for evaluating Flt-1 and its ligands as therapeutic targets for promoting revascularization of ischemic tissues and for blocking uncontrolled angiogenesis and inflammation in cancer, arthritis, atherosclerosis, and retinal ischemia.

REFERENCES

1. FERRARA, N. 2001. Role of vascular endothelial growth factor in regulation of physiological angiogenesis. Am. J. Physiol. Cell Physiol. **280:** C1358–C1366.
2. ISNER, J.M. 2002. Myocardial gene therapy. Nature **415:** 234–239.
3. POST, M.J. *et al.* 2001. Therapeutic angiogenesis in cardiology using protein formulations. Cardiovasc. Res. **49:** 522–531.
4. CARMELIET, P. & R.K. JAIN. 2000. Angiogenesis in cancer and other diseases. Nature **407:** 249–257.

5. VEIKKOLA, T. et al. 2000. Regulation of angiogenesis via vascular endothelial growth factor receptors. Cancer Res. **60:** 203–212.
6. HIRATSUKA, S., et al. 1998. Flt-1 lacking the tyrosine kinase domain is sufficient for normal development and angiogenesis in mice. Proc. Natl. Acad. Sci. USA **95:** 9349–9354.
7. SHIBUYA, M. 2001. Structure and function of VEGF/VEGF-receptor system involved in angiogenesis. Cell Struct. Funct. **26:** 25–35.
8. PERSICO, M.G., V. VINCENTI & T. DIPALMA. 1999. Structure, expression and receptor-binding properties of placenta growth factor (PlGF). Curr. Topics Microbiol. Immunol. **237:** 31–40.
9. FONG, G.H. et al. 1995. Role of the Flt-1 receptor tyrosine kinase in regulating the assembly of vascular endothelium. Nature **376:** 66–70.
10. PARK, J.E. et al. 1994. Placenta growth factor. Potentiation of vascular endothelial growth factor bioactivity, in vitro and in vivo, and high affinity binding to Flt-1 but not to Flk-1/KDR. J. Biol. Chem. **269:** 25646–25654.
11. CARMELIET, P. et al. 2001. Synergism between vascular endothelial growth factor and placental growth factor contributes to angiogenesis and plasma extravasation in pathological conditions. Nat. Med. **7:** 575–583.
12. HIRATSUKA, S., et al. 2001. Involvement of Flt-1 tyrosine kinase (vascular endothelial growth factor receptor-1) in pathological angiogenesis. Cancer Res. **61:** 1207–1213.
13. WENG, D.E. & N. USMAN. 2001. Angiozyme: a novel angiogenesis inhibitor. Curr. Oncol. Rep. **3:** 141–146.
14. CARMELIET, P. et al. 1999. Impaired myocardial angiogenesis and ischemic cardiomyopathy in mice lacking the vascular endothelial growth factor isoforms VEGF164 and VEGF188. Nat. Med. **5:** 495–502.
15. SIMONS, M. et al. 2000. Clinical trials in coronary angiogenesis: issues, problems, consensus: An expert panel summary. Circulation **102:** E73–E86.
16. PETTERSSON, A. et al. 2000. Heterogeneity of the angiogenic response induced in different normal adult tissues by vascular permeability factor/vascular endothelial growth factor. Lab. Invest. **80:** 99–115.
17. DVORAK, H.F. et al. 1979. Fibrin gel investment associated with line 1 and line 10 solid tumor growth, angiogenesis, and fibroplasia in guinea pigs. Role of cellular immunity, myofibroblasts, microvascular damage, and infarction in line 1 tumor regression. J. Natl. Cancer Inst. **62:** 1459–1472.
18. CARMELIET, P. 2000. Mechanisms of angiogenesis and arteriogenesis. Nat. Med. **6:** 389–395.
19. VAN ROYEN, N. et al. 2001. Stimulation of arteriogenesis; a new concept for the treatment of arterial occlusive disease. Cardiovasc. Res. **49:** 543–553.
20. HOEFER, I.E. et al. 2001. Time course of arteriogenesis following femoral artery occlusion in the rabbit. Cardiovasc. Res. **49:** 609–617.
21. ARRAS, M. et al. 1998. Monocyte activation in angiogenesis and collateral growth in the rabbit hindlimb. J. Clin. Invest. **101:** 40–50.
22. CARMELIET, P. & E.M. CONWAY. 2001. Growing better blood vessels. Nat. Biotechnol. **19:** 1019–1020.
23. ISHIDA, A. et al. 2001. Expression of vascular endothelial growth factor receptors in smooth muscle cells. J. Cell Physiol. **188:** 359–368.
24. LAZAROUS, D.F. et al. 1996. Comparative effects of basic fibroblast growth factor and vascular endothelial growth factor on coronary collateral development and the arterial response to injury. Circulation **94:** 1074–1082.

25. MCLEOD, D.S. et al. 2002. Localization of VEGF receptor-2 (KDR/Flk-1) and effects of blocking it in oxygen-induced retinopathy. Invest. Ophthalmol. Vis. Sci. **43:** 474–482.
26. BOTTOMLEY, M.J. et al. 2000. Placenta growth factor (PlGF) induces vascular endothelial growth factor (VEGF) secretion from mononuclear cells and is co-expressed with VEGF in synovial fluid. Clin. Exp. Immunol. **119:** 182–188.
27. LYDEN, D. et al. 2001. Impaired recruitment of bone-marrow-derived endothelial and hematopoietic precursor cells blocks tumor angiogenesis and growth. Nat. Med. **7:** 1194–1201.
28. MOULTON, K.S. 2001. Plaque angiogenesis and atherosclerosis. Curr. Atheroscler. Rep. **3:** 225–233.
29. CELLETTI, F.L. et al. 2001. Effect of human recombinant vascular endothelial growth factor165 on progression of atherosclerotic plaque. J. Am. Coll. Cardiol. **37:** 2126–2130.
30. RIESSEN, R. et al. 1998. Immunolocalization of thrombospondin-1 in human atherosclerotic and restenotic arteries. Am. Heart J. **135:** 357–364.
31. LEE, D.M. & M.E. WEINBLATT. 2001. Rheumatoid arthritis. Lancet **358:** 903–911.
32. WEBER, K.T. et al. 1994. Collagen network of the myocardium: function, structural remodeling and regulatory mechanisms. J. Mol. Cell. Cardiol. **26:** 279–292.
33. MIOTLA, J. et al. 2000. Treatment with soluble VEGF receptor reduces disease severity in murine collagen-induced arthritis. Lab. Invest. **80:** 1195–1205.
34. SONE, H. et al. 2001. Neutralization of vascular endothelial growth factor prevents collagen-induced arthritis and ameliorates established disease in mice. Biochem. Biophys. Res. Commun. **281:** 562–568.
35. LUROSS, J.A. & N.A. WILLIAMS. 2001. The genetic and immunopathological processes underlying collagen-induced arthritis. Immunology **103:** 407–416.
36. SAWANO, A. et al. 2001. Flt-1, vascular endothelial growth factor receptor 1, is a novel cell surface marker for the lineage of monocyte-macrophages in humans. Blood **97:** 785–791.
37. CLAUSS, M. 1998. Functions of the VEGF receptor-1 (FLT-1) in the vasculature. Trends Cardiovasc. Med. **8:** 241–245.
38. BRENCHLEY, P.E. 2000. Angiogenesis in inflammatory joint disease: a target for therapeutic intervention. Clin. Exp. Immunol. **121:** 426–429.
39. FOLKMAN, J. 2001. Angiogenesis-dependent diseases. Semin. Oncol. **28:** 536–542.
40. LIBBY, P. 2001. What have we learned about the biology of atherosclerosis? The role of inflammation. Am. J. Cardiol. **88:** 3J–6J.
41. BURLEIGH, M.E., et al. 2002. Cyclooxygenase-2 promotes early atherosclerotic lesion formation in LDL receptor-deficient mice. Circulation **105:** 1816–1823.
42. LUTGENS, E. et al. 2000. Both early and delayed anti-CD40L antibody treatment induces a stable plaque phenotype. Proc. Natl. Acad. Sci. USA **97:** 7464–7469.
43. CARMELIET, P. 2000. VEGF gene therapy: stimulating angiogenesis or angioma-genesis? Nat. Med. **6:** 1102–1103.

Insights into the Molecular Pathogenesis and Targeted Treatment of Lymphedema

ANNE SAARISTO, MARIKA J. KARKKAINEN, AND KARI ALITALO

Molecular/Cancer Biology Laboratory, Biomedicum, University of Helsinki, Helsinki, Finland

ABSTRACT: Abnormal function of the lymphatic vessels is associated with a variety of diseases, such as tumor metastasis and lymphedema. The development of strategies for local and controlled induction or inhibition of lymphangiogenesis would thus be of major importance for the treatment of such diseases. Two growth factors, vascular endothelial growth factor C (VEGF-C) and D (VEGF-D), have been found to be important in the proper formation and maintenance of the lymphatic network, through their receptor VEGFR-3. In patients with lymphedema, heterozygous inactivation of VEGFR-3 leads to primary lymphedema due to defective lymphatic drainage in the limbs. We have shown that VEGF-C gene transfer to the skin of mice with lymphedema induces regeneration of the cutaneous lymphatic vessel network. However, as is the case with VEGF, high levels of VEGF-C cause blood vessel growth and leakiness, resulting in tissue edema. Strategies to avoid these side-effects have also been developed. This new field of reseach has important implications for the development of new therapies for human lymphedema.

KEYWORDS: gene therapy; adenovirus; AAV; VEGFR-3; missense mutation; mouse model

INTRODUCTION

Impairment of the lymphatic function is involved in various diseases characterized by inadequate transport of interstitial fluid, edema, impaired immunity, and fibrosis.[1] Moreover, the lymphatic vessels serve as an important route for tumor metastasis. Until very recently, little was known about the molecular mechanisms behind these diseases. The first gene mutations causing human lymphedema have been found; the first lymphatic markers and

---

Address for correspondence: Kari Alitalo, M.D., Ph.D., Molecular/Cancer Biology Laboratory, Biomedicum, University of Helsinki, P.O.B. 63 (Haartmaninkatu 8), SF-00014 Helsinki, Finland. Voice: +358-9-191 25511; fax: +358-9-191 25510.
Kari.Alitalo@Helsinki.FI

lymphangiogenic growth factors have been identified; and several mouse models have facilitated the development of therapeutic applications for lymphedema. Animal models of tumor lymphangiogenesis have also been used to test various lymphangiogenic inhibitors that would inhibit tumor metastasis.[2]

The novel findings allow the modulation of the lymphangiogenic process and specific targeting of the lymphatic endothelium. We now have evidence for the first effective lymphangiogenic gene therapy. The approach has shown very promising results in an experimental animal model. Here we provide insights into the mechanisms of physiological and pathological lymphagiogenesis and discuss the use of VEGF-C gene transfer as a pro-lymphangiogenic agent in lymphedema therapy.

THE FORMATION OF THE LYMPHATIC VESSELS DURING EMBRYOGENESIS

During embryogenesis, the formation of new blood vessels occurs via two processes, vasculogenesis and angiogenesis (FIG. 1). Vasculogenesis involves the *de novo* differentiation of endothelial cells from mesoderm-derived precursor cells called angioblasts, which cluster and reorganize to form capillary-like tubes.[3] Once the primary vascular plexus is formed, new capillaries form by sprouting or by splitting (intussusception) from pre-existing vessels.[4]

Lymphatic vessels develop shortly after blood vessels, at around midgestation. There are two theories about the origin of the lymphatic vessels. A century ago, Sabin proposed that the primitive lymph sacs originate by endothelial cell budding from the pre-existing embryonic veins (FIG. 1).[5] The peripheral lymphatic system would then spread from these primary lymphatic sacs by sprouting. An alternative model suggests that the initial lymph sacs arise in the mesenchyme from precursor cells, independent of veins, and that the connection to the venous system is formed later in development.[6] Recently, two lymphatic-specific markers, VEGFR-3 and Prox1, have been show to be expressed in the endothelium that lines the suggested lymphatic sacs in mouse embryos, supporting Sabin's theory of lymphatic development.[7–10] In addition, it has been shown, in a quail-chick chimera model, that the mesodermal lymphangioblasts participate in the development of the lymphatic system, supporting the theory that the peripheral lymphatic vessels develop by multiple mechanisms.[11,12] Whether the lymphangioblasts can participate in the lymphangiogenesis of mammalian tissues and adults must still be analyzed. In theory, potential lymphatic precursor cells could be utilized in various pro-lymphangiogenic approaches.

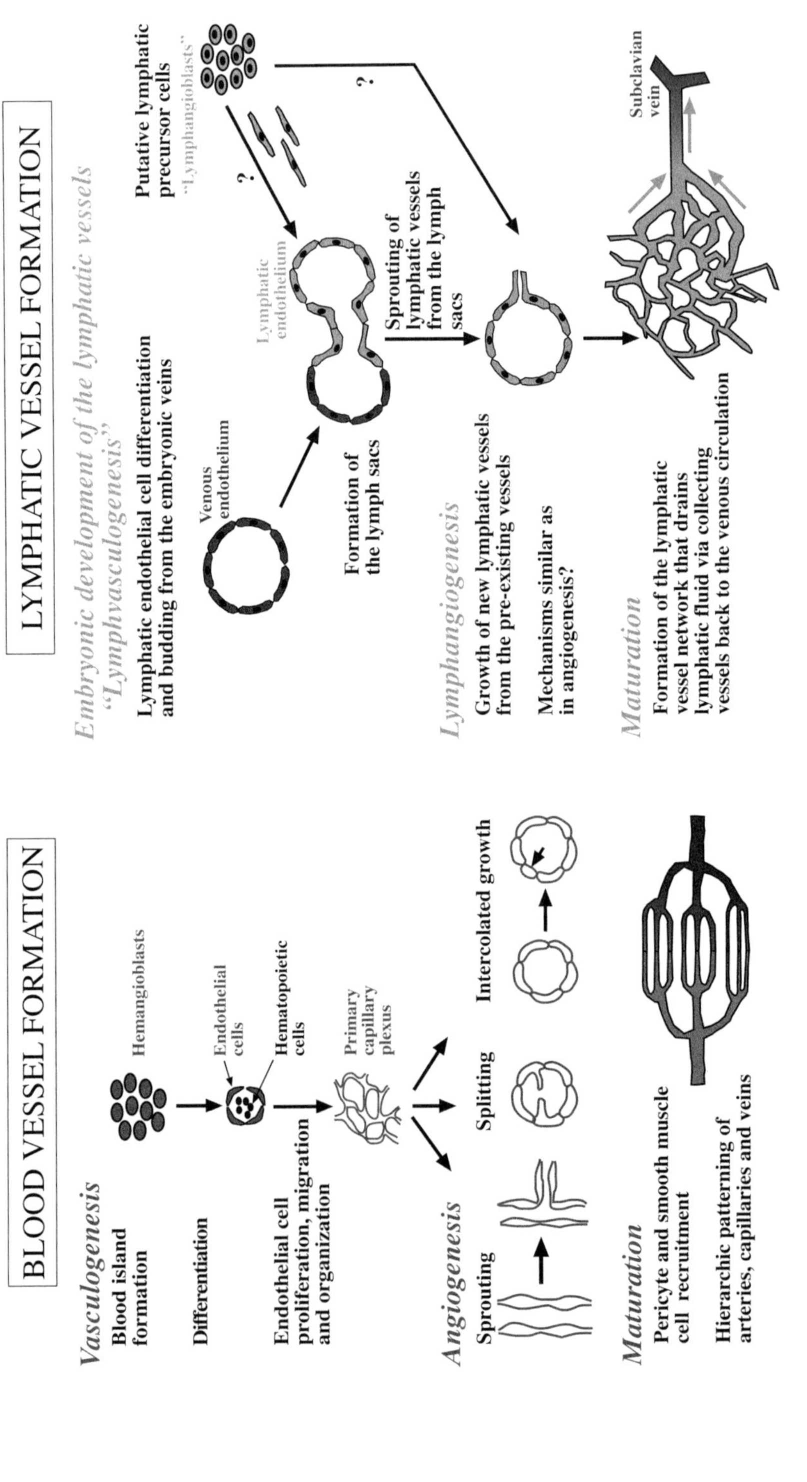

FIGURE 1. *See following page for legend.*

VEGF-C/VEGFR-3 IN LYMPHANGIOGENESIS

VEGF-C and VEGF-D are two growth factors capable of inducing growth of new lymphatic vessels *in vivo*. These factors belong to the larger VEGF family of growth factors, which also includes VEGF, placenta growth factor (PlGF), and VEGF-B. VEGF-C and VEGF-D activate the endothelial cell-specific tyrosine kinase receptors VEGFR-2 and VEGFR-3.[13,14] VEGF-C is mitogenic towards lymphatic endothelial cells and shows a selective lymphangiogenic response in differentiated avian chorioallantoic membrane.[15] Accordingly, overexpression of VEGF-C or VEGF-D in transgenic mice induces the development of a hyperplastic lymphatic vessel network.[16–18] Recent data also suggest that a VEGFR-3-specific mutant of VEGF-C (VEGF-C156S) is lymphangiogenic when overexpressed in the skin of transgenic mice.[18,19] Conversely, inhibition of lymphatic growth was obtained when VEGF-C/VEGF-D binding to their receptors was blocked by a soluble form of VEGFR-3 in a similar transgenic mouse model.[20]

VEGF-C and VEGF-D mediate their effects via the VEGFR-3 tyrosine kinase receptor that is expressed predominantly in the lymphatic endothelial cells lining the inner surface of lymphatic vessels.[8,14,21–23] During embryogenesis, VEGFR-3 is first expressed in blood vascular endothelial cells (ECs).[8] Accordingly, mice deficient in the *Vegfr3* gene show abnormal remodeling of the primary vascular plexus and die at E9.5.[7] However, during further development, *Vegfr3* is abundant in lymphatic endothelium and down-regulated elsewhere.[7,8] The results discussed in the preceding text, and the expression patterns of VEGF-C and VEGFR-3, suggest that lymphangiogenesis is induced by VEGF-C/VEGF-D and mediated via VEGFR-3. Interestingly, neuropilin-2, a receptor for various VEGFs on venous endothelium, and semaphorins on neural cells, also acts as a co-receptor for VEGF-C in some lymphatic vessels.[24,25] It has also been shown previously that neuropilin-2–deficient mice show absence or severe reduction of small lymphatic vessels during development.[26]

Proteolytic cleavage is an important regulator of the receptor binding and, thus, the biological activity, of VEGF-C and VEGF-D.[13,27] Partially processed forms of VEGF-C and VEGF-D are able to activate VEGFR-3, while the fully processed short forms are also potent stimulators of VEGFR-2, expressed in both blood and lymphatic vessel endothelia.[28] VEGF-C has

---

FIGURE 1. Mechanisms of angiogenesis and lymphangiogenesis. The blood vessels form by successive steps of vasculogenesis and angiogenesis to form a hierarchic system with arteries, capillaries, and veins. The lymphatic vessels originate by differentiating and budding or sprouting from embryonic veins, but lymphangioblasts may also participate in the formation of lymphatic sacs and lymphangiogenesis. See text for details.

been also shown to increase blood vessel permeability *in vivo*,[13,29,30] and high levels of VEGF-C cause enlargement and tortuosity of the veins and venules that have been shown to express VEGFR-2.[30] Neutralizing VEGFR-2 antibodies have been shown to block the vascular permeability effect of VEGF-C in tumor models.[29]

The physiological role of VEGF-C as an inducer of blood vessel permeability is also supported by the finding that VEGFR-3 is expressed in some fenestrated blood vessel endothelia, such as in some endocrine glands and in the respiratory epithelium of the nasal cavity.[31,32] Thus, VEGF-C expressed by the neuroendocrine cells and by the respiratory epithelial cells may regulate the permeability of the nearby VEGFR-3- and VEGFR-2-positive fenestrated blood vessels. In tumors, VEGF-C may stimulate lymphangiogenesis and lymphatic metastasis and act as an adjunctive stimulator of tumor angiogenesis (reviewed by Karkkainen *et al.*[33] and Pepper[34]). Similarly, VEGF-D overexpression in a tumor xenograft model resulted in both angiogenic and lymphangiogenic responses.[35] However, at least in tumor models, VEGF-D did not cause hyperpermeability of the tumor blood vessels.[35]

OTHER MOLECULES FOR TARGETING THE LYMPHATIC VESSELS

Increasing numbers of lymphatic markers and growth factors that target the lymphatic vessels have recently been found (TABLE 1). The transcription factor Prox1 is required for the programming of the lymphatic endothelial cell budding and sprouting from veins during embryogenesis.[9,10] Prox1 is expressed in a variety of different cell types, but, among endothelial cells, its expression is restricted to the lymphatic endothelium.[9]

The lymphatic endothelial hyaluronan receptor (LYVE-1) is a CD44 homologue that was identified as a cell-surface protein specific for lymphatic endothelial cells and macrophages.[36,37] However, like VEGFR-3, LYVE-1 is also expressed in liver sinusoidal endothelial cells.[38] LYVE-1 binds hyaluronan (HA), an abundant tissue glycosaminoglycan, that plays a role in the maintenance of tissue integrity and cell migration.[37] In the lymphatic vessels, LYVE-1 seems to play role in transporting HA across the lymphatic vessel wall.[37] Further studies should reveal whether LYVE-1-HA interactions are involved in leukocyte migration and tumor metastasis.

Another recently described novel marker for lymphatic endothelium is podoplanin.[39] In addition to lymphatic endothelium, this surface glycoprotein is expressed in several other cell types, including kidney podocytes, osteoblastic cells, and lung alveolar cells.[40] The function of podoplanin in the lymphatic endothelium is not known. Detailed comparison of the expression patterns of Prox1, VEGFR-3, LYVE-1, and podoplanin in different endothelia and in the tumor vasculature requires further study.

TABLE 1. Lymphatic vessel markers

Marker	Protein Class	Biological Effect
VEGFR-3	Receptor tyrosine kinase on ECs	Lymphangiogenesis; survival of LECs
LYVE-1	Receptor for extracellular matrix glycosaminoglycan	Transport of hyaluronan from tissues to lymph nodes
Podoplanin	Integral membrane mucoprotein	Unknown
Prox1	Homeobox transcription factor	Involved in budding and sprouting of lymphatic vessels during development
β-Chemokine receptor D6	Chemokine receptor in the afferent lymphatics	Leukocyte recirculation
Macrophage mannose receptor	Receptor in macrophages, lymphoid organs, lymphatic endothelial cells, perivascular microglia and glomerular mesangial cells	Phagocytosis of microbes; viral endocytosis
VEGFR-3	Receptor for extracellular matrix glycosaminoglycan	Cell–cell adhesion of LECs

Other markers of lymphatic endothelial cells include macrophage mannose receptor and desmoplakin.[41–43] In addition, there are now also several molecules reported to be important in lymphatic development, such as the transcription factor Net, β-chemokine receptor D6, integrin α9β1, and angiopoietin (Ang)-2.[44–47] Recently, methods have been published to isolate and culture lymphatic endothelial cells separately from the blood vascular endothelial cells.[48,49] Further studies of the gene and protein expression patterns of these two isolated cell populations should result in the discovery of new lymphatic-specific markers (TABLE 2).[50]

LYMPHEDEMA

Lymphatic vessels play a key role in the immune response to various antigens and in maintaining fluid homeostasis in the body. Blockage of the lymphatic drainage or an abnormal development of the superficial lymphatic vessels leads to lymphedema, which is characterized by disfiguring and disabling swelling of the extremities. The molecular pathogenesis of various lymphedema phenotypes has been unclear, but recent reports indicate several chromosomal regions and genes that are involved in the development of lymphedema. Congenital hereditary lymphedema (Milroy's disease) was previously linked to the *VEGFR3* region on distal chromosome 5q, and inacti-

TABLE 2. Blood and lymphatic endothelial cells display distinct transcription profiles

	Blood vascular EC	Lymphatic EC
Adhesion molecules	**Integrin α5** Integrin β5 **ICAM-1***, ICAM-2 **N-cadherin*** Selectin P, selectin E* CD44*	**Integrin α9*** **Integrin α1**
Cytoskeletal proteins	**Vinculin** **Claudin 7*** Actin, alpha 2 Profilin 2	**Desmoplakin I and II*** **Adducin gamma** **Plakoglobin** Alpha-actinin-2 associated LIM protein*
ECM proteins	**Collagens 8A1*, 6A1*, 1A2*** Laminin, gamma 2*, alpha 5 Versican* Proteoglycan 1	**Matrix Gla protein***
ECM modulation	**MMP-1**, MMP-14 uPA* Plasminogen activator inhibitor I Cathepsin C	**TIMP-3**
Cytokines, chemokines, and receptors	**IL-8***, IL-6*	IL-7*
	Monocyte chemotactic protein 1 UFO/axl* **CXCR4** CCRL2/CKRX* IL-4 receptor	SDF-1b*
Total	167 genes	135 genes

NOTE: Examples of different classes of proteins are shown. Genes in bold were confirmed by Northern blotting or immunofluorescence; those marked with an asterisk (*) were specifically expressed in only one of the two cell lineages (EC: endothelial cell; ECM: extracellular matrix). Modified from reference 50, with permission from Oxford University Press.

vating VEGFR-3 missense mutations were found to be causative for the disease (FIG. 2).[51–55] While mutations that inhibit the biological activity of VEGFR-3 are one cause of primary lymphedema, there are several families with Milroy's disease and other lymphedema syndromes, which involve other genetic loci (TABLE 3). For example, *FOXC2* gene mutations are causative for lymphedema–distichiasis,[56] and characterization of other genes involved in

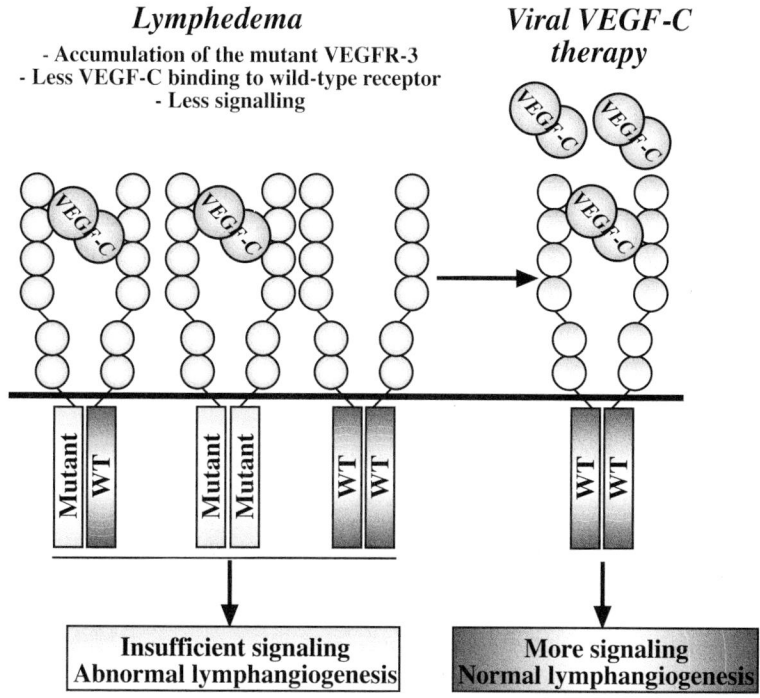

FIGURE 2. VEGFR-3 in pathogenesis and treatment of primary lymphedema. Heterozygous inactivation of the VEGFR-3 tyrosine kinase function is one cause of congenital lymphedema. In VEGF-C lymphedema therapy, excess of VEGF-C ligand can normalize VEGFR-3 signaling and stimulate the growth of new lymphatic vessels.

TABLE 3. Genetic alterations in lymphedema syndromes

Lymphedema Syndrome	MIM[a]	Age at Onset	Gene Loci	Gene	References
Milroy's disease	153100	Congenital	5q34-q35	VEGFR3	51–55
Lymphedema–distichiasis	153400	Puberty	16q24.3	*FOXC2*	56, 72, 73
Cholestasis–lymphedema syndrome	214900	Puberty	15q	Not known	74
Turner syndrome	–	Congenital	Xp11.2-p22.1	Not known	75
Noonan syndrome	163950	Congenital	12q24.1	PTPN11 (SHP-2)	76–78

[a]Mendelian inheritance in man (http://www3.ncbi.nlm.nih.gov/omim/).

the development of lymphedema syndromes will give us more insight into the molecular mechanisms of lymphedema.

MOUSE MODELS FOR LYMPHEDEMA

Several experimental models of secondary lymphedema have been developed recently, including lymphedema in the mouse tail,[57] hindlimb lymphedema in rats,[58,59] and lymphedema in the rabbit ear.[60] In addition, two mouse lines showing phenotypes of primary lymphedema have been described. In the Chy model, an inactivating *Vegfr3* mutation results in persistent hypoplasia of the superficial lymphatic vessels. On the contrary, the subserosal lymphatic vessels are enlarged in these mice, and this leads to formation of chylous ascites after birth.[25] In another model, overexpression of a soluble VEGFR-3 in transgenic mice competes for VEGF-C/VEGF-D binding with the endogenous recptor.[20] This led to regression of the extant lymphatic vessels in several organs and resulting lymphedema. However, the lymphatic vessels regenerated during later postnatal development in most organs, except in the skin. As in human lymphedema patients, both of these mouse models show swelling of the limbs because of a hypoplastic/aplastic cutaneous lymphatic vessel network. Thus, these genetic mouse models provide us new tools to develop and test new therapies for lymphatic dysfunction.

VEGF-C THERAPY FOR LYMPHEDEMA

Development of strategies for local and controlled induction of lymphangiogenesis is of major importance for the treatment of both primary and secondary lymphedema. The discovery of specific genes and signaling cascades involved in the regulation of the lymphatic vessel function and in the pathogenesis of lymphatic dysfunctions have established basis for the development for new targeted treatments for these diseases.

Previously, pro-angiogenic gene therapy in humans, developed first by Dr. Jeffrey Isner, has shown promise in the treatment of cardiovascular ischemic diseases.[61] Angiogenesis has been stimulated by overexpression of VEGF or various fibroblast growth factors, using several gene transfer vectors (reviewed by Ferrara and Alitalo[62]). Research on the field of pro-lymphagiogenesis began only recently, and we still have a lot to learn from the previous pro-angiogenesis studies.

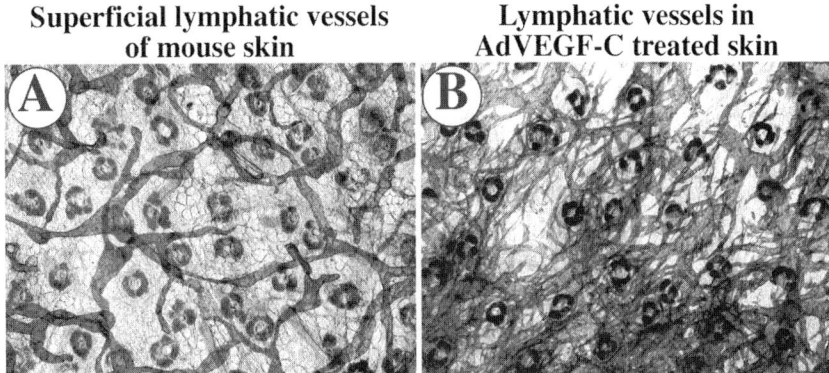

FIGURE 3. Lymphangiogenesis after AdVEGF-C gene transfer. VEGFR-3 whole-mount immunohistochemical staining shows the **(A)** normal and **(B)** VEGF-C-activated, sprouting lymphatic vessels in mouse skin.

Vectors for Gene Therapy

Intradermal adenovirus and adeno-associated virus (AAV)-mediated VEGF-C gene transfer into the mouse skin has been shown to result in a strong lymphangiogenic response (FIG. 3).[30,63] Recombinant adenoviruses are efficient and commonly used gene transfer vectors, but their major limitation is that the transgene expression is lost within a month, due to an immune response to the remaining viral proteins of the vector. Whereas the adenoviral gene transfer yields only short-term expression, AAVs yield transgene expression that may last for over a year.[64] AAVs are non-pathogenic human viruses, which do not elicit an inflammatory reaction or a cytotoxic immune response, and infect both dividing and non-dividing cells of several organs (reviewed by Monahan and Samulski[65]). Recombinant AAVs have also been shown to result in long-term gene expression in humans. The lower transgene expression levels and slower kinetics of the AAV-VEGF-C infection result in a more controlled lymphangiogenesis than the acute high-level expression obtained by the use of adenoviral vectors. AAV could thus be more suitable for lymphangiogenic therapy in humans.

VEGF-C Gene Therapy in Chy Lymphedema Mice

Based on studies of virally delivered VEGF-C, we have recently also tested gene transfer of VEGF-C in the skin of the Chy lymphedema mice with inherited partial loss of VEGFR-3 activity. As in the human patients, some VEGFR-3 activity remains in the Chy mice due to the functional wild-type

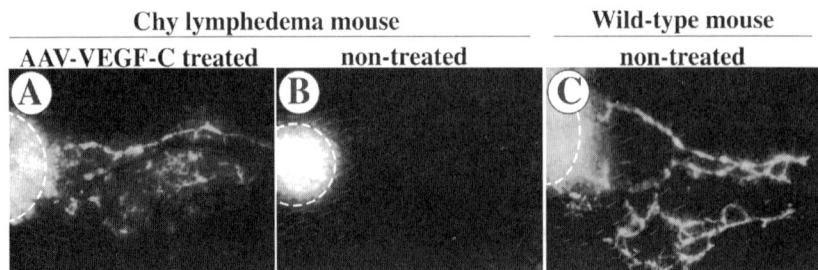

FIGURE 4. Formatiom of functional lymphatic vessels in the skin of the Chy lymphedema mice in response to VEGF-C gene therapy. (**A**) Fluorescent microlymphography reveals formation of a functional lymphatic vessel network in the skin of the Chy mice after AAV-VEGF-C administration. Note the comparisons with untreated control (**B**) and wild-type mouse skin (**C**).

allele, and we therefore wanted to explore the possibility of using VEGF-C ligand overexpression as a therapeutic tool for primary lymphedema.[25] Interestingly, both adeno- and AAV-mediated VEGF-C overexpression resulted in formation of a functional lymphatic vessel network in the skin of the Chy mice, which normally lack the superficial lymphatic vessels (FIG. 4). In AAV-VEGF-C-infected Chy mice that have been followed for up to 8 months after infection, functional skin lymphatic vessels are still present.[28]

VEGFR-3 signaling plays role in lymphatic endothelial cell survival,[20,48] and recent studies have shown that the majority of the lymphatic vessels formed in response to adenoviral VEGF-C expression regress when the adenovirus is no longer active.[28] In contrast, the AAV-mediated gene transfer results into long-lasting transgene expression, and therefore the VEGF-C-induced lymphatic vessels were also sustained in the Chy mice.[28] Whether a single recombinant AAV infection is sufficient to induce life-long transgene expression is not known. If the problems with the gene delivery can be solved, VEGF-C or VEGF-D gene therapy could be applicable in the treatment of both heriditary and non-heriditary regional forms of lymphedema caused by trauma, surgery, or filariasis.

Is It Possible to Develop Lymphagiogenic Gene Therapy without Blood Vascular Side-Effects?

As discussed above, high levels of VEGF-C lead to blood vascular effects such as increased vessel leakiness in the skin.[13,30] In the treatment of human lymphatic dysfunction and edema, this is an undesirable side-effect. Overexpression of Ang1 has been shown to block the permeability effect of both VEGF and VEGF-C.[30,66,67] Importantly, Ang1 does not block the lymphangiogenic effect of VEGF-C;[30] therefore, combination therapy with Ang1 and VEGF-C might be used.

Proteolytically processed forms of VEGF-C/VEGF-D bind to VEGFR-2, and the blood vascular effects are presumably mediated via this signaling pathway.[19] One possibility to avoid the permeability effects of VEGF-C would be to use the VEGFR-3-specific mutant form of this growth factor (VEGF-C156S).[19] Interestingly, the corresponding mutation in rat VEGF-D abolished binding to both VEGFR-2 and VEGFR-3.[68] The lymphangiogenic properties of VEGF-C156S has been recently tested in different animal models, including the Chy mouse model.[18,25,28] These reports show that both transgenic and viral overexpression of VEGF-C156S is sufficient, but not as effective as VEGF-C for inducing the growth of functional lymphatic vessel network. It was also confirmed that viral VEGF-C156S lacks the blood vascular side-effects of the native VEGF-C in the skin.[25,28] The other possible concern in the therapeutic use of VEGF-C is that it might stimulate tumor lymphangiogenesis and lymphatic metastasis of dormant tumors.[2] In particular, this risk should be carefully evaluated before the therapeutic use of VEGF-C is entertained in cancer patients with postoperative secondary lymphedema.

FUTURE ASPECTS OF LYMPHEDEMA THERAPY

As illustrated in the foregoing discussion, vectors that result in long-term gene expression are probably needed for prolymphangiogenic therapy. In the future, narrowing the target-cell range of the vectors or gene expression could be advantageous. Therapeutic applications may include targeting of gene expression to the desired endothelial cells of the specific tissue.[69] This goal may be achieved by targeting viral entry to the desired cells only. The approach is facilitated by the discovery of new lymphatic and blood endothelial-cell- or pericyte-specific markers; thus, distinct targeting molecules could be used.[69] Alternatively, endothelial-cell-specific promoters, such as those of the *TIE1*, *TEK*, *KDR*, *ICAM1*, or endoglin genes, could be used to achieve endothelial-specific expression of the therapeutic gene. The lymphatic endothelial-cell-specific promoters, such as the *VEGFR3* regulatory region, could also be used to target the lymphatic endothelium.[70]

The first lymphangiogenic growth factors have been discovered, but the molecular control of the formation of a patterned hierarchy of lymphatic vessels is still largely unknown. Since both capillaries and collecting vessels are needed for the lymphatic drainage function, more attention should be paid to the development of strategies to specifically target the different types of lymphatic vessels. In the pro-angiogenesis studies, it has been shown that although VEGF is a potent inducer of neovasculature, the newly formed vessels are immature, tortuous, and leaky, and they often lack the perivascular support structures.[71] This also applies to the newly formed collecting lymphatic vessels; thus, and the characteristics of the new lymphatic vessels should be studied in detail. Designing an optimal pro-lymphangiogenic therapy is facilitated by finding that endothelial cell receptor expression varies among different lymphatic vessels.[28] Thus, growth factor combinations could be utilized to create a more functional, hierarchical lymphatic network. These issues will be of major importance when designing molecular therapies for human lymphedema.

CONCLUSIONS

Understanding the molecular mechanisms of different forms of lymphedema and the discovery of lymphangiogenic growth factors and lymphatic markers now allows us to develop targeted therapies for lymphatic dysfunction. In mouse models, pro-lymphangiogenic VEGF-C and VEGF-C156S therapies seem very promising. The next steps will include the optimization of the therapeutic approaches and clinical trials in patients. These studies will finally show the therapeutic potential of the pro-lymphangiogenic therapy.

ACKNOWLEDGMENTS

The authors have been supported by grants from the Finnish Cultural Foundation, the Central Finland Central Hospital (Paavo and Eila Salonen Foundation), the Finnish Cancer Organization, the Emil Aaltonen Foundation, the Ida Montini Foundation, the Paulo Foundation, the Research and Science Foundation of Farmos, the Academy of Finland, the Novo Nordisk Foundation, the National Institutes of Health (Grant HD37243), and the European Union (Biomed Grant PL963380).

REFERENCES

1. ROCKSON, S.G. 2001. Lymphedema. Am. J. Med. **110:** 288–295.
2. KARPANEN, T. & K. ALITALO. 2001. Lymphatic vessels as targets of tumor therapy. J. Exp. Med. **194:** F34–F42.
3. RISAU, W. & I. FLAMME. 1995. Vasculogenesis. Annu. Rev. Cell Dev. Biol. **11:** 73–91
4. RISAU, W. 1997. Mechanisms of angiogenesis. Nature **386:** 671–674.
5. SABIN, F.R. 1902. On the origin of the lymphatic system from the veins and the development of the lymph hearts and thoracic duct in the pig. Am. J. Anat. **1:** 367–391.

6. HUNTINGTON, G.S. & C.F.W. MCCLURE. 1908. The anatomy and development of the jugular lymph sac in the domestic cat (*Felis domestica*). Anat. Rec. **2:** 1–19.
7. DUMONT, D.J., L. JUSSILA, J. TAIPALE, *et al.* 1998. Cardiovascular failure in mouse embryos deficient in VEGF receptor-3. Science **282:** 946–949.
8. KAIPAINEN, A., J. KORHONEN, T. MUSTONEN, *et al.* 1995. Expression of the *fms*-like tyrosine kinase FLT4 gene becomes restricted to lymphatic endothelium during development. Proc. Natl Acad. Sci. USA **92:** 3566–3570.
9. WIGLE, J.T. & G. OLIVER. 1999. Prox1 function is required for the development of the murine lymphatic system. Cell **9:** 769–778.
10. WIGLE, J.T., N. HARVEY, M. DETMAR, *et al.* 2002. An essential role for Prox1 in the induction of the lymphatic endothelial cell phenotype. EMBO J. **21:** 1505–1513.
11. SCHNEIDER, M., K. OTHMAN-HASSAN, B. CHRIST, *et al.* 1999. Lymphangioblasts in the avian wing bud. Dev. Dyn. **216:** 311–319.
12. WILTING, J., M. SCHNEIDER, M. PAPOUTSI, *et al.* 2000. An avian model for studies of embryonic lymphangiogenesis. Lymphology **33:** 81–94.
13. JOUKOV, V., T. SORSA, V. KUMAR, *et al.* 1997. Proteolytic processing regulates receptor specificity and activity of VEGF-C. EMBO J. **16:** 3898–3911.
14. ACHEN, M.G., M. JELTSCH, E. KUKK, *et al.* 1998. Vascular endothelial growth factor D (VEGF-D) is a ligand for the tyrosine kinases VEGF receptor 2 (Flk1) and VEGF receptor 3 (Flt4). Proc. Natl Acad. Sci. USA **95:** 548–553.
15. OH, S.-J., M.M. JELTSCH, R. BIRKENHAGER, *et al.* 1997. VEGF and VEGF-C: specific induction of angiogenesis and lymphangiogenesis in the differentiated avian chorioallantoic membrane. Dev. Biol. **188:** 96–109.
16. JELTSCH, M., A. KAIPAINEN, V. JOUKOV, *et al.* 1997. Hyperplasia of lymphatic vessels in VEGF-C transgenic mice. Science **276:** 1423–1425.
17. MANDRIOTA, S.J., L. JUSSILA, M. JELTSCH, *et al.* 2001. Vascular endothelial growth factor-C-mediated lymphangiogenesis promotes tumour metastasis. EMBO J. **20:** 672–682.
18. VEIKKOLA, T., L. JUSSILA, T. MAKINEN, *et al.* 2001. Signalling via vascular endothelial growth factor receptor-3 is sufficient for lymphangiogenesis in transgenic mice. EMBO J. **6:** 1223–1231.
19. JOUKOV, V., V. KUMAR, T. SORSA, *et al.* 1998. A recombinant mutant vascular endothelial growth factor-C that has lost vascular endothelial growth factor receptor-2 binding, activation, and vascular permeability activities. J. Biol. Chem. **273:** 6599–6602.
20. MÄKINEN, T., L. JUSSILA, T. VEIKKOLA, *et al.* 2001. Inhibition of lymphangiogenesis with resulting lymphedema in transgenic mice expressing soluble VEGF receptor-3. Nat. Med. **7:** 199–205.
21. GALLAND, F., A. KARAMYSHEVA, M.-J. PEBUSQUE, *et al.* 1993. The *FLT4* gene encodes a transmembrane tyrosine kinase related to the vascular endothelial growth factor receptor. Oncogene **8:** 1233–1240.
22. PAJUSOLA, K., O. APRELIKOVA, J. KORHONEN, *et al.* 1992. FLT4 receptor tyrosine kinase contains seven immunoglobulin-like loops and is expressed in multiple human tissues and cell lines. Cancer Res. **52:** 5738–5743.
23. JOUKOV, V., K. PAJUSOLA, A. KAIPAINEN, *et al.* 1996. A novel vascular endothelial growth factor, VEGF-C, is a ligand for the Flt4 (VEGFR-3) and KDR (VEGFR-2) receptor tyrosine kinases. EMBO J. **15:** 290–298.

24. HERZOG, Y., C. KALCHEIM, N. KAHANE, *et al.* 2001. Differential expression of neuropilin-1 and neuropilin-2 in arteries and veins. Mech. Dev. **109:** 115–119.
25. KARKKAINEN, M.J., A. SAARISTO, L. JUSSILA, *et al.* 2001. A model for gene therapy of human hereditary lymphedema. Proc. Natl Acad. Sci. USA **98:** 12677–12682.
26. YUAN, L., D. MOYON, L. PARDANAUD, *et al.* 2002. Abnormal lymphatic vessel development in neuropilin-2 mutant mice. Development **129:** 4797–4806.
27. STACKER, S.A., K. STENVERS, C. CAESAR, *et al.* 1999. Biosynthesis of vascular endothelial growth factor-D involves proteolytic processing which generates non-covalent homodimers. J. Biol. Chem. **274:** 32127–32136.
28. SAARISTO, A., T. VEIKKOLA, T. TAMMELA, *et al.* 2002. Lymphangiogenic gene therapy without blood vascular side-effects. Manuscript submitted for publication.
29. KADAMBI, A., C.M. CARREIRA, C. YUN, *et al.* 2001. Vascular endothelial growth factor (VEGF)-C differentially affects tumor vascular function and leukocyte recruitment: Role of VEGF-receptor 2 and host VEGF-A. Cancer Res. **61:** 2404–2408.
30. SAARISTO, A., T. VEIKKOLA, B. ENHOLM, *et al.* 2002. Adenoviral VEGF-C overexpression induces blood vessel enlargement, tortuosity and leakiness, but no sprouting angiogenesis in the skin or mucous membranes. FASEB J. **16:** 1041–1049.
31. PARTANEN, T.A., J. AROLA, A. SAARISTO, *et al.* 2000. VEGF-C and VEGF-D expression in neuroendocrine cells and their receptor, VEGFR-3, in fenestrated blood vessels in human tissues. FASEB J. **14:** 2087–2096.
32. SAARISTO, A., T.A. PARTANEN, J. AROLA, *et al.* 2000. Vascular endothelial growth factor-C and its receptor VEGFR-3 in the nasal mucosa and in nasopharyngeal tumors. Am. J. Pathol. **157:** 7–14.
33. KARKKAINEN, M.J., T. MÄKINEN, K. ALITALO. 2002. Lymphatic endothelium: a new frontier of metastasis research. Nat. Cell Biol. **4:** E2–E5.
34. PEPPER, M.S. 2001. Lymphangiogenesis and tumor metastasis: myth or reality? Clin. Cancer Res. **7:** 462–468.
35. STACKER, S.A., C. CAESAR, M.E. BALDWIN, *et al.* 2001. VEGF-D promotes the metastatic spread of tumor cells via the lymphatics. Nat. Med. **7:** 186–191.
36. BANERJI, S., J. NI, S.-X. WANG, *et al.* 1999. LYVE-1, a new homologue of the CD44 glycoprotein, is a lymph-specific receptor for hyaluronan. J. Cell Biol. **144:** 789–801.
37. JACKSON, D.G., R. PREVO, S. CLASPER, *et al.* 2001. LYVE-1, the lymphatic system and tumor lymphangiogenesis. Trends Immunol. **22:** 317–321.
38. CARREIRA, C.M., S.M. NASSER, E. DI TOMASO *et al.* 2001. LYVE-1 is not restricted to the lymph vessels: Expression in normal liver blood sinusoids and down-regulation in human liver in cancer and cirrhosis. Cancer Res. **61:** 8079–8084.
39. BREITENEDER-GELEFF, S., A. SOLEIMAN, H. KOWALSKI, *et al.* 1999. Angiosarcomas express mixed endothelial phenotypes of blood and lymphatic capillaries: podoplanin as a specific marker for lymphatic endothelium. Am. J. Pathol. **154:** 385–394.
40. WETTERWALD, A., W. HOFFSTETTER, M.G. CECCHINI *et al.* 1996. Characterization and cloning of the E11 antigen, a marker expressed by rat osteoblasts and osteocytes. Bone **18:** 125–132.

41. LINEHAN, S.A., L. MARTINEZ-POMARES, P.D. STAHL, et al. 1999. Mannose receptor and its putative ligands in normal murine lymphoid and nonlymphoid organs: In situ expression of mannose receptor by selected macrophages, endothelial cells, perivascular microglia, and mesangial cells, but not dendritic cells. J. Exp. Med. **189:** 1961–1972.
42. IRJALA, H., E.L. JOHANSSON, R. GRENMAN, et al. 2001. Mannose receptor is a novel ligand for L-selectin and mediates lymphocyte binding to lymphatic endothelium. J. Exp. Med. **194:** 1033–1042.
43. SCHMELZ, M. & W.W. FRANKE. 1993. Complexus adhaerentes, a new group of desmoplakin-containing junctions in endothelial cells: the syndesmos connecting retothelial cells of lymph nodes. Eur. J. Cell Biol. **61:** 274–289.
44. AYADI, A., H. ZHENG, P. SOBIESZCZUK, et al. 2001. Net-targeted mutant mice develop a vascular phenotype and up-regulate egr-1. EMBO J. **20:** 5139–5152.
45. NIBBS, R.J.B., E. KRIEHUBER, P.D. PONATH, et al. 2001. The b-chemokine receptor D6 is expressed by lymphatic endothelium and a subset of vascular tumors. Am. J. Pathol. **158:** 867–877.
46. HUANG, X.Z., J.F. WU, R. FERRANDO, et al. 2000. Fatal bilateral chylothorax in mice lacking the integrin a9b1. Mol. Cell. Biol. **20:** 5208–5215.
47. GALE, N.W., G. THURSTON, S.F. HACKETT, et al. 2002. Angiopoietin-2 is required for postnatal angiogenesis and lymphatic patterning, and only the latter role is rescued by angiopoietin-1. Dev. Cell **3:** 411–423.
48. MÄKINEN, T., T. VEIKKOLA, S. MUSTJOKI, et al. 2001. Isolated lymphatic endothelial cells transduce growth, survival and migratory signals via the VEGF-C receptor VEGFR-3. EMBO J. **20:** 4762–4773.
49. KRIEHUBER, E., S. BREITENEDER-GELEFF, M. GROEGER, et al. 2001. Isolation and characterization of dermal lymphatic and blood endothelial cells reveal stable and functionally specialized cell lineages. J. Exp. Med. **194:** 797–808.
50. PETROVA, T.V., T. MAKINEN, T.P. MAKELA, et al. 2002. Lymphatic endothelial reprogramming of vascular endothelial cells by the Prox-1 homeobox transcription factor. EMBO J. **21:** 4593–4599.
51. FERRELL, R.E., K.L. LEVINSON, J.H. ESMAN, et al. 1998. Hereditary lymphedema: evidence for linkage and genetic heterogeneity. Hum. Mol. Genet. **7:** 2073–2078.
52. WITTE, M.H., R. ERICKSON, M. BERNAS, et al. 1998. Phenotypic and genotypic heterogeneity in familial Milroy lymphedema. Lymphology **31:** 145–155.
53. EVANS, A.L., G. BRICE, V. SOTIROVA, et al. 1999. Mapping of primary congenital lymphedema to the 5q35.3 region. Am. J. Hum. Genet. **64:** 547–555.
54. KARKKAINEN, M.J., R.E. FERRELL, E.C. LAWRENCE, et al. 2000. Missense mutations interfere with VEGFR-3 signalling in primary lymphoedema. Nat. Genet. **25:** 153–159.
55. IRRTHUM, A., M.J. KARKKAINEN, K. DEVRIENDT, et al. 2000. Congenital hereditary lymphedema caused by a mutation that inactivates VEGFR3 tyrosine kinase. Am. J. Hum. Genet. **67:** 295–301.
56. FANG, J., S.L. DAGENAIS, R.P. ERICKSON, et al. 2000. Mutations in FOXC2 (MFH-1), a forkhead family transcription factor, are responsible for the hereditary lymphedema-distichiasis syndrome. Am. J. Hum. Genet. **67:** 1382–1388.
57. SWARTZ, M.A., A. KAIPAINEN, P.A. NETTI, et al. 1999. Mechanics of interstitial-lymphatic fluid transport: theoretical foundation and experimental validation. J. Biomech. **32:** 1297–1307.
58. LEE-DONALDSON, L., M.H. WITTE, M. BERNAS, et al. 1999. Refinement of a rodent model of peripheral lymphedema. Lymphology **32:** 111–117.

59. KRIEDERMAN, B., T. MYLOYDE, M. BERNAS, *et al.* 2002. Limb volume reduction after physical treatment by compression and/or massage in a rodent model of peripheral lymphedema. Lymphology **35:** 23–27.
60. SZUBA, A., M. SKOBE, M.J. KARKKAINEN, *et al.* Therapeutic lymphangiogenesis with VEGF-C. FASEB J. In press.
61. ISNER, J.M. 2002. Myocardial gene therapy. Nature **415:** 234–243.
62. FERRARA, N. & K. ALITALO. 1999. Clinical applications of angiogenic growth factors and their inhibitors. Nat. Med. **5:** 1359–1364.
63. ENHOLM, B., T. KARPANEN, M. JELTSCH, *et al.* 2001. Adenoviral expression of vascular endothelial growth factor-C induces lymphangiogenesis in the skin. Circ. Res. **88:** 623–629.
64. DALY, T.M., K.K. OHLEMILLER, M.S. ROBERTS, *et al.* 2001. Prevention of systemic clinical disease in MPS VII mice following AAV-mediated neonatal gene transfer. Gene Ther. **8:** 1291–1298.
65. MONAHAN, P.E. & R.J. SAMULSKI. 2000. Adeno-associated virus vectors for gene therapy: more pros than cons? Mol. Med. Today **6:** 433–440.
66. THURSTON, G., C. SURI, K. SMITH, *et al.* 1999. Leakage-resistant blood vessels in mice transgenically overexpressing angiopoietin-1. Science **286:** 2511–2514.
67. THURSTON, G., J.S. RUDGE, E. IOFFE, *et al.* 2000. Angiopoietin-1 protects the adult vasculature against plasma leakage. Nat. Med. **6:** 460–463.
68. KIRKIN, V., R. MAZITSCHEK, J. KRISHNAN, *et al.* 2001. Characterization of indolinones which preferentially inhibit VEGF-C- and VEGF-D-induced activation of VEGFR-3 rather than VEGFR-2. Eur. J. Biochem. **268:** 5530–5540.
69. RUOSLAHTI, E. & D. RAJOTTE. 2000. An address system in the vasculature of normal tissues and tumors. Ann. Rev. Immunol. **18:** 813–827.
70. ILJIN, K., M.J. KARKKAINEN, E.C. LAWRENCE, *et al.* 2001. VEGFR3 gene structure, regulatory region and sequence polymorphisms. FASEB J. **15:** 1028–1036.
71. CARMELIET, P. 2000. VEGF gene therapy: stimulating angiogenesis or angioma-genesis? Nat. Med. **6:** 1102–1103.
72. FINEGOLD, D.N., M.A. KIMAK, E.C. LAWRENCE, *et al.* 2001. Truncating mutations in FOXC2 cause multiple lymphedema syndromes. Hum. Mol. Genet. **10:** 1185–1189.
73. BELL, R., G. BRICE, A.H. CHILD, *et al.* 2001. Analysis of lymphoedema-distichiasis families for FOXC2 mutations reveals small insertions and deletions throughout the gene. Hum. Genet. **108:** 546–551.
74. BULL, L.N., E. ROCHE, E.J. SONG, *et al.* 2000. Mapping of the locus for cholestasis-lymphedema syndrome (Aagenaes syndrome) to a 6.6-cM interval on chromosome 15q. Am. J. Hum. Genet. **67:** 994–999.
75. ZINN, A.R., V.S. TONK, Z. CHEN, *et al.* 1998. Evidence for a Turner syndrome locus or loci at Xp11.2-p22.1. Am. J. Hum. Genet. **63:** 1757–1766.
76. TARTAGLIA, M., E.L. MEHLER, R. GOLDBERG, *et al.* 2001. Mutations in PTPN11, encoding the protein tyrosine phosphatase SHP-2, cause Noonan syndrome. Nat. Genet. **29:** 465–468.
77. WHITE, S.W. 1984. Lymphedema in Noonan's syndrome. Int. J. Dermatol. **2:** 656–657.
78. WITT, D.R., H.E. HOYME, J. ZONANA, *et al.* 1987. Lymphedema in Noonan syndrome: clues to pathogenesis and prenatal diagnosis and review of the literature. Am. J. Med. Genet. **27:** 841–856.

Lymphangiogenesis

New Mechanisms

LYNN CHANG, ARJA KAIPAINEN, AND JUDAH FOLKMAN

Department of Surgery, Harvard Medical School, and Surgical Research Laboratory, Children's Hospital, Boston, Massachusetts 02115, USA

> ABSTRACT: A mouse model has been developed to study lymphangiogenesis dissociated from angiogenesis. bFGF implanted in a mouse cornea at a concentration below the threshold to induce angiogenesis potently induces lymphangiogenesis. This model has permitted a study of cellular and molecular mechanisms of lymphangiogenesis.
>
> KEYWORDS: angiogenesis; lymphangiogenesis; bFGF; VEGF-A; VEGF-D; angiogenesis inhibitor; lymphangiogenesis inhibitor; lymphatic endothelial cells; blood vessel endothelial cells; murine cornea neovascularization

INTRODUCTION

The process of angiogenesis, new growth of capillary blood vessels, is now recognized as an important control point in cancer.[1,2] The hypothesis that tumors are angiogenesis-dependent, has been confirmed by a variety of methods,[1] and more recently by genetic experiments.[3–7] As a result, the microvascular endothelial cell recruited by a tumor has become an important second target in cancer therapy. It is a genetically stable target, in contrast to the tumor cell. Angiogenesis inhibitors developed to suppress tumor angiogenesis are emerging as a new class of drugs, several of which are in clinical trial. Therefore, it has become feasible to propose that treating the endothelial cell in a tumor bed, or treating both the endothelial cell and the tumor cell may be more effective than treating the cancer cell alone.

While the microvascular endothelial cell has been extensively studied since its first successful *in vitro* culture almost three decades ago,[8,9] the lymphatic endothelial cell is less well understood. The relation of lymphatic endothelial cell growth to tumor growth or tumor progression remains problematic, mainly because of the paucity of bioassays and other methods.

Address for correspondence: Judah Folkman, M.D., Children's Hospital, Hunnewell 103, 300 Longwood Avenue, Boston, MA 02115. Voice: 617-355-7661; fax: 617-739-5891.
judah.folkman@tch.harvard.edu

Therefore, it has been difficult to address certain important questions about the connection of lymphatics to tumor biology. For example: (1) Is lymphangiogenesis under different regulatory mechanisms than angiogenesis? (2) Why do many tumors appear devoid of lymphatics (except for engorged lymphatics at the periphery of a tumor), while some tumors appear to contain a dense internal lymphatic network? (3) Is the relative deficiency or absence of lymphatics in some tumors the result of mechanical compression of lymphatic vessels or, instead, of humoral suppression of lymphatic growth? (4) Do all angiogenesis inhibitors also inhibit lymphangiognesis, or only some, or none? (5) Are some tumors capable of suppressing lympangiogenesis at a remote site, analogous to the suppression of angiogenesis in remote metastases by a primary tumor.[10,11]

These questions have long been discussed in our laboratory. but they could not be easily studied, because it was not possible to clearly dissociate angiogenesis from lymphangiogenesis in an *in vivo* system. Lynn Chang and Arja Kaipainen in our laboratory have solved this problem by modifying the mouse cornea neovascularization model.

LYMPHANGIOGENESIS IN THE MOUSE CORNEA

In 1974, when small tumors of approximately 1 mm^3 were first implanted in a pocket made in the rabbit cornea,[12] capillary blood vessels grew from the limbal edge of the cornea over a distance of approximately 2 millimeters and reached the implanted tumor by 7–10 days. Once the tumor was neovascularized, rapid tumor growth followed. The purpose of this experiment was to challenge the dogma prevalent among scientists at that time that tumors did not need new blood vessels and could not induce them, especially from a distance, by secreting a diffusible angiogenic factor. Later, when implantable sustained-release polymeric pellets were developed,[13] they were loaded with soluble tumor extracts and subsequently with a purified angiogenic protein,[14] bFGF (basic fibroblast growth factor). Robert Auerbach demonstrated that tumors or angiogenic pellets could be implanted in the mouse cornea where new capillary blood vessels arising from the limbal edge of the cornea could traverse approximately 1 millimeter of avascular cornea and reach the edge of the pocket in approximately 5–6 days.[15] In our laboratory lyophilized bFGF, 80 ng, is dispersed in a sustained-release pellet made of polyhydroxyethyl methacrylate (polyHEMA) and allowed to dry before inserting it into a mouse corneal pocket under general anesthesia.[16,17] The pellet is implanted off-center in the cornea, so that it is approximately 1 millimeter from the limbal edge at 6 o'clock. The neovascularization induced by an implanted polymeric pellet containing 80 ng bFGF develops over a period of 6 days, the average time required for new microvessels to reach the bFGF pellet across the 1 millimeter distance. This method has been widely used by many laboratories for more than a decade. However, it has not been possible to observe lymphatics.

Lynn Chang, a postdoctoral fellow in our laboratory, stained a neovascularized cornea with antibody to CD-31 conjugated to fluorescein isothiocyanate (FITC), which stains blood vessels and lymphatic vessels (FIG. 1). Blind-ending lymphatic sprouts could be visually distinguished from blood vessel loops. He observed that although blood vessels had entered the cornea and had converged on the bFGF pellet only over a narrow sector where the bFGF was closest to the limbus, new lymphatic vessels were entering the cornea from its entire circumference, including the limbal edge of cornea, which was at a 3-fold greater distance from the bFGF pellet. This pattern suggested that lymphatic vessels were being stimulated at a concentration of bFGF *below* the bFGF concentration necessary for blood vessels. Furthermore, when the bFGF pellet was moved to the center of the cornea, there was a significant decrease in blood vessel growth and an increase in over-all lymphatic growth. When the concentration of bFGF in the pellet was reduced to 12.5 ng, and this "low concentration" bFGF pellet was moved back to its original off-center position, *lymphatic vessels now reached the pellet, but blood vessels did not.* For the first time lymphatic vessels could be dissociated from blood vessels in this bioassay.

Lynn Chang now teamed up with Arja Kaipainen, also a postdoctoral fellow who had come to this laboratory two years earlier from Kari Alitalo's laboratory in Helsinki. Kaipainen's Ph.D. thesis was on VEGFR3 (vascular endothelial growth factor receptor 3), which she found to be the first marker for the lymphatic system.[18] In mice with a 12.5-ng bFGF pellet in the cornea, Chang and Kaipainen administered intravenous lectin conjugated to fluorescein isothiocyanate for 3 minutes. All blood vessels were stained green, but there was not sufficient time for any dye to enter lymphatics (FIG. 2A). The mice were then euthanized and the corneas were removed and stained overnight with anti-CD31 antibody conjugated to phycoerythrin. Blood vessels and lymphatic vessels all stained red (FIGURE 2B). When the images were merged, blood vessels were yellow and lymphatics were red (FIG. 2C).

The molecular evidence that these were lymphatic vessels is provided in detail in a forthcoming paper by Chang *et al.* (submitted for publication).

To determine which cell types in the cornea express VEGF-A, C, and D, mRNA *in situ* hybridization was performed. These results indicate that VEGF-D was expressed in the cornea, probably from resident keratocytes. (Chang *et al.*, submitted for publication).

To address the question of whether or not any tumors can suppress lymphangiogenesis by a humoral mechanism as opposed to the simple mechanical compression in a tumor resulting from high interstitial pressure, different tumors were implanted in mice in the subcutaneous dorsal position. When the tumors had reached 1000 to 2000 mm^3 a bFGF pellet was implanted in the

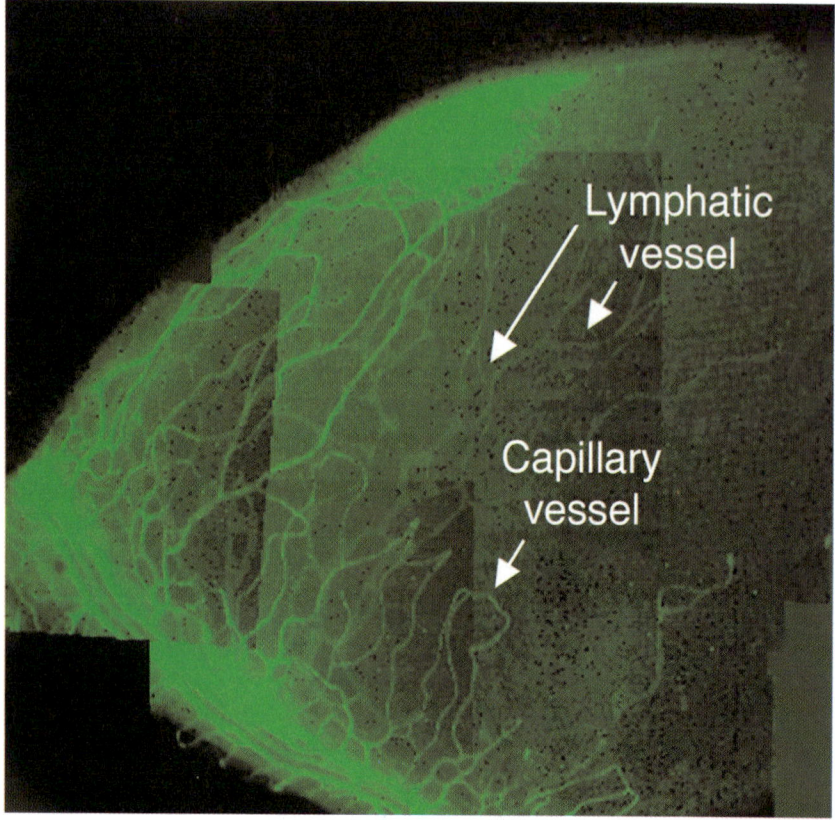

FIGURE 1. High-dose bFGF induces angiogenesis and lymphangiogenesis. We first demonstrated in December 1999 that bFGF induced the growth of lymphatic vessels in the cornea. A pellet containing 80 ng of bFGF was implanted in the corneal stroma. When the whole cornea and limbus were removed and incubated with anti-PECAM-FITC, all blood vessels and lymphatic vessels stained green. The pellet is green because of autofluorescence.

cornea. A fibrosarcoma (FIG. 3A) almost completely inhibited lymphangiogenesis, while a melanoma (FIG. 3B) had no effect and permitted full and intense lymphangiogenesis in the cornea. This experiment suggests that fibrosarcoma may generate a circulating inhibitor of lymphangiogenesis, possibly by a similar mechanism as angiostatin and endostatin are generated (i.e., by enzymatic cleavage of a cryptic internal fragment from a larger protein). This putative inhibitor has not yet been purified at this writing. Nor is it known whether the circulating "lymphangiogenesis inhibitor" is another function of angiostatin or endostatin or, instead, a novel protein with lymphangiostatic activity.

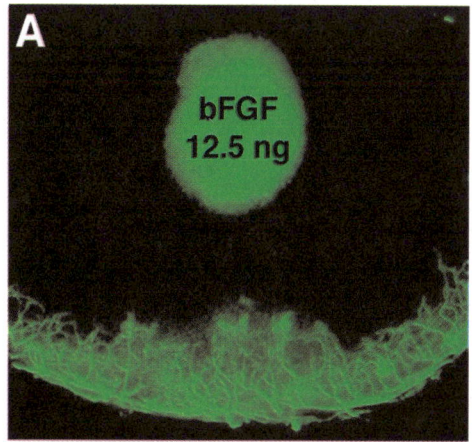
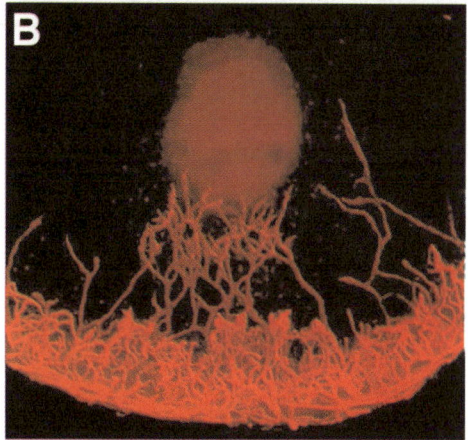
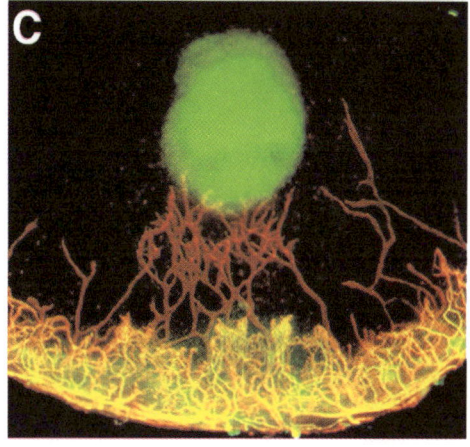

FIGURE 2. Modification of the corneal pocket assay results in selective stimulation of lymphangiogenesis. The concentration of bFGF in the pellet was lowered to 12.5 ng, and the pellet was implanted farther away from the limbus. After a week, blood vessels were perfused with i.v. lectin-FITC (green, **A**), and then whole cornea and limbus were stained with anti-PECAM-PE (red, **B**). When both images were merged blood vessels appeared yellow, and lymphatic vessels were red (**C**).

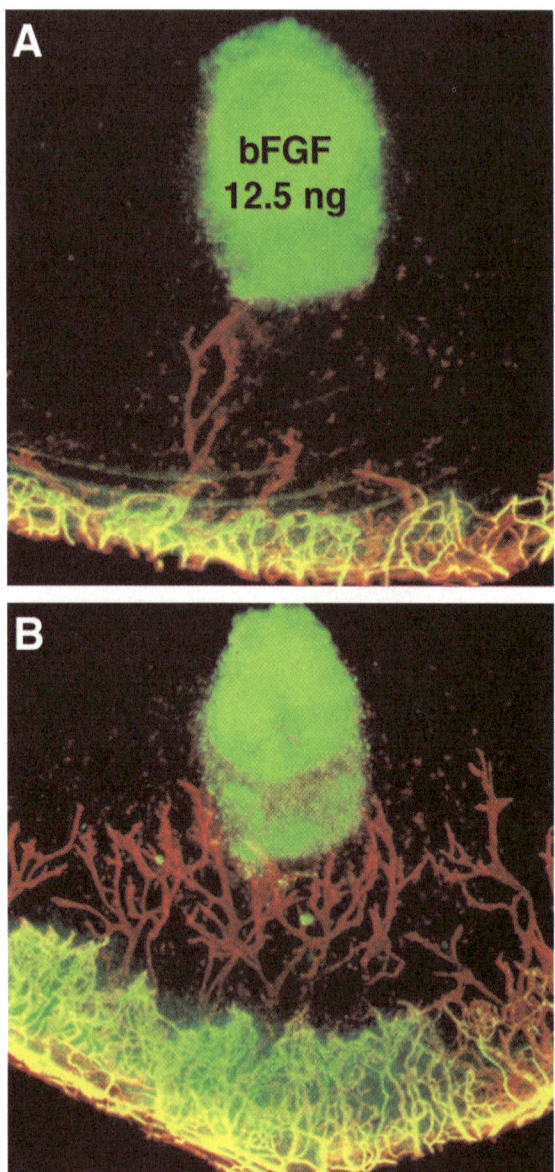

FIGURE 3. Corneal lymphangiogenesis is inhibited by a distant tumor. Tumors were growing subcutaneously on the backs of mice when low-dose bFGF pellets were implanted into the cornea. T241 fibrosarcoma inhibited bFGF-stimulated lymphangiogenesis and angiogenesis (**A**), whereas B16 melanoma did not inhibit either (**B**).

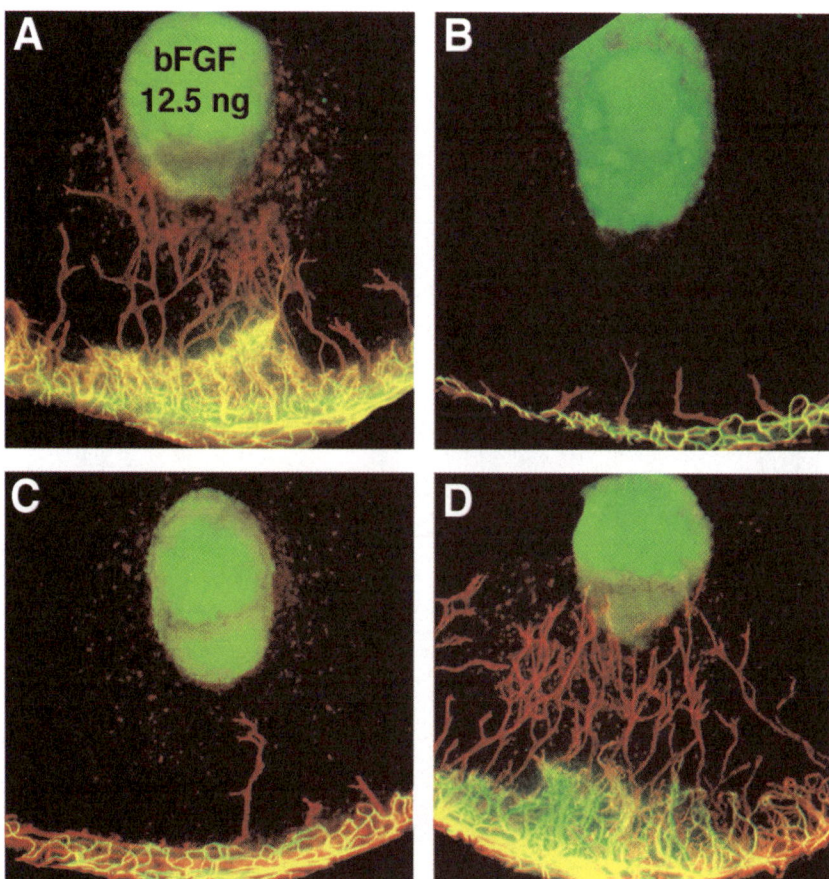

FIGURE 4. Corneal lymphangiogenesis is inhibited by COX-2 inhibitors, but not by thalidomide. (**A**) bFGF-stimulated lymphangiogenesis in a control mouse receiving oral vehicle only. Oral administration of rofecoxib (20 mg/kg/day) and celecoxib (60 mg/kg/day) potently prevented 12.5-ng-bFGF-stimulated lymphangiogenesis (**B,C**). Thalidomide (200 mg/kg/day) failed to inhibit lymphangiogenesis (**D**).

The ability to dissociate lymphangiogenesis from angiogenesis permitted the development of a simple bioassay to determine which angiogenesis inhibitors, if any, can also inhibit lymphangiogenesis. This bioassay was validated by the demonstration that oral administration of either of two cyclooxygenase-2 inhibitors, Vioxx (rofecoxib) (FIG. 4A AND B) and Celebrex (celecoxib) (FIG. 4C) both potently inhibited lymphangiogenesis and angiogenesis. A single intraperitoneal dose of cyclophosphamide also inhibited lymphangiogenesis. In contrast, thalidomide (FIG. 4D) did not inhibit lymphangiogenesis.

DISCUSSION

These studies show that low concentrations of bFGF in the cornea preferentially induce lymphangiogenesis over angiogenesis. Lymphangiogenesis can now be studied in isolation from angiogenesis.

Our results favor a hypothesis that the relative paucity of lymphatics in certain tumors may be due to an inhibitor of lymphangiogenesis generated by the tumor. The concept of mechanical compression of lymphatics is a less tenable concept. Furthermore, there may be at least two categories of tumors: those that can inhibit lymphangiogenesis and those that cannot. A tumor capable of generating a circulating inhibitor of lymphangiogenesis may suppress metastasis to lymph nodes, which may explain why some human sarcomas do not metastasize by lymphatics.

Therefore, the classic patterns described in textbooks (that for some tumors metastases are mainly by hematogenous spread) could possibly be explained by the fact that the primary tumor may dictate the route of metastatic spread. For example, it has previously been reported that when breast cancer cells (which were already highly angiogenic) were transfected with VEGF-C, tumors showed lymphangiogenesis and increased metastases.[19] In contrast, other tumors that are known to spread by hematogenous and lymphatic pathways may be unable to suppress remote lymphatic metastasis, e.g., melanoma.

These results also provide a practical method of determining for each angiogenesis inhibitor whether lymphangiogenesis is also inhibited. It remains to be seen whether a pure lymphangiogenesis inhibitor that does not inhibit angiogenesis will be discovered.

After the bFGF pellet becomes depleted (e.g., at approximately 2 weeks), it will no longer stimulate new blood vessels or new lymphatic vessels. In fact, new capillary blood vessels in the cornea usually begin to regress about two weeks after the bFGF pellet has become depleted. In contrast, the lymphatic vessels remain intact for up at least 6 months, and in many mice for up to one year. The explanation of this difference is not clear, especially because the lymphatic vessels are very thin-walled and lack pericytes.

It is possible that these studies will also lead to an increased understanding of the development of lymphatic malformations, and to potential novel therapeutic approaches.

Now that lymphangiogenesis can be studied in isolation, progress in understanding this vascular process may proceed as rapidly as angiogenesis research has advanced in the past decade.

[NOTE ADDED IN PROOF: Supported in part by the Daland Clinical Investigator Fellowship (to L.C.) from the American Philosophical Society; Grant RO1CA64481 from the NCI, NIH, DHHS (to J.F.); and the CaP CURE Foundation.]

REFERENCES

1. FOLKMAN, J. 2000. Tumour angiogenesis. *In* Cancer Medicine, 5th ed. J.F. Holland *et al.*, Eds. :32–152. B.C. Decker. Ontario, Canada.

2. FOLKMAN, J. 2001. Angiogenesis. In Harrison's Textbook of Internal Medicine, 15th ed. E. Braunwald et al. Eds. :517–530. McGraw-Hill. New York.
3. ARBISER, J.L., M.A. MOSES, C.A. FERNANDEZ, et al. 1997. Oncogenic H-ras stimulates tumor angiogenesis by two distinct pathways. Proc. Natl. Acad. Sci. USA **94**: 861–866.
4. CHIN, L., A. TAM, J. POMERANTZ, et al. 1999. Essential role for oncogene Ras in tumour maintenance. Nature **400**: 468–472.
5. STREIT, M., L. RICCARDI, P. VELASCO, et al. 1999. Thrombospondin-2: a potent endogenous inhibitor of tumour growth and angiogenesis. Proc. Natl. Acad. Sci. USA **96**: 14888–14893.
6. LYDEN, D., A.Z. YOUNG, D. ZAGZAG, et al. 1999. Id1 and Id3 are required for neurogenesis, angiogenesis and vascularization of tumour xenografts. Nature **401**: 670–677.
7. LYDEN, D., K. HATTORI & S. DIAS. 2001. Impaired recruitment of bone-marrow-derived endothelial and hematopoietic precursor cells blocks tumor angiogenesis and growth. Nature Med. **7**: 1194–1201.
8. GIMBRONE, M.A., JR., R.S. COTRAN & J. FOLKMAN. 1973. Endothelial regeneration: studies with human endothelial cells in culture. Ser. Haemat. **VI**: 453–455.
9. JAFFE, E.A., R.L. NACHMAN, C.G. BECKER & R.C. MINICK. 1972. Culture of human endothelial cells derived from human cord umbilical cord veins. Circulation. **46**: II–252.
10. O'REILLY, M.S., L. HOLMGREN, Y. SHING, et al. 1994. Angiostatin: a novel angiogenesis inhibitor that mediates the suppression of metastases by a Lewis lung carcinoma. Cell. **79**: 315–328.
11. HOLMGREN, L., M.S. O'REILLY & J. FOLKMAN. 1995. Dormancy of micrometastases: balanced proliferation and apoptosis in the presence of angiogenesis suppression. Nature Med. **1**: 149–153.
12. GIMBRONE, M.A., JR, R.S. COTRAN, S.B. LEAPMAN & J. FOLKMAN. 1974. Tumor growth and neovascularization: an experimental model using rabbit cornea. J. Natl. Cancer Inst. **52**: 413–427.
13. LANGER, R. & J. FOLKMAN. 1976. Polymers for sustained release of proteins and other macromolecules. Nature **263**: 797–800.
14. SHING, Y., J. FOLKMAN, R. SULLIVAN, et al. 1984. Heparin-affinity: purification of a tumor-derived capillary endothelial cell growth factor. Science **223**: 1296–1299.
15. MUTHUKKARUPPAN, V.R. & R. AUERBACH. 1979. Angiogenesis in the mouse cornea. Science **205**: 1416–1418.
16. FOLKMAN, J. 1985. Angiogenesis and its inhibitors. In Important Advances in Oncology. V.T. DeVita, Jr., S. Hellman & S.A. Rosenberg, Eds. :42–62. J.B. Lippincott. Philadelphia, PA.
17. KENYON, B.M., E.E. VOEST, C.C. CHEN, et al. 1996. A model of angiogenesis in the mouse cornea. Invest. Ophthal. Vis. Sci. **37**: 1625–1632.
18. KAIPAINEN, A., J. KORHONEN, T. MUSTONEN, et al. Expression of the fms-like tyrosine kinase 4 gene becomes restricted to lymphatic endothelium during development. Proc Natl. Acad Sci USA **92**: 3566–3570.
19. SKOBE, M., T. HAWIGHORST, D.G. JACKSON, et al. 2001. Induction of tumor lymphangiogenesis by VEGF-C promotes breast cancer metastasis. Nature Med. **7**: 192–198.

Lymphatic Vessel Activation in Cancer

MELANIE CASSELLA AND MIHAELA SKOBE

Derald H. Ruttenberg Cancer Center, Mount Sinai School of Medicine, New York, New York 10029, USA

ABSTRACT: Metastasis of most cancers occurs primarily through the lymphatic system, and the extent of lymph node involvement is the most important prognostic indicator. While the importance of the lymphatic system as a pathway for metastasis has been well recognized, there is very little information available about the mechanisms by which tumor cells interact with the lymphatics. Recently, production of the lymphangiogenic factor VEGF-C has been detected in tumors, and the significance of VEGF-C-mediated lymphangiogenesis for tumor metastasis has been demonstrated. Increased lymphatic vessel density has been found associated with certain tumors. The mechanisms by which tumor cells gain access to and enter lymphatic vessels are critical issues that need to be addressed in the future. In contrast to the prevailing view that has assigned to the lymphatic system a passive role in the metastatic process, our results indicate the importance of lymphatic vessel activation in tumor dissemination.

KEYWORDS: lymphangiogenesis; lymphatic endothelium; cancer metastasis; VEGF-C; VEGF receptor

INTRODUCTION

The metastatic spread of tumor cells is a major cause of death in cancer patients. The lymphatic system is the primary pathway of metastasis for most human cancers, and the extent of lymph node involvement is a key prognostic factor for the patient's outcome. In spite of this, most experimental work addressing tumor dissemination has focused on hematogenous spread.[1,2] In fact, the ability of tumor cells to induce angiogenesis is considered a prerequisite for tumor growth, invasion, and successful metastasis, and the angio-

Address for correspondence: Mihaela Skobe, Ph.D., Derald H. Ruttenberg Cancer Center, Mount Sinai School of Medicine, One Gustave L. Levy Place, Box 1130, New York, NY 10029. Voice: 212-659-5570; fax: 212-987-2240.
mihaela.skobe@mssm.edu

genic switch is recognized as one of the key events in tumorigenesis. In contrast, very little effort has been directed towards understanding the molecular regulation of lymphatic vessel formation and function, and the major issues regarding the role of lymphatic vessels in tumor growth and metastasis remain unresolved.[3] We are now beginning to gain insight into the function of lymphatic vessels in tumor progression, and it remains to be determined whether lymphangiogenesis and/or activation of the lymphatic system is an integral part of tumorigenesis in humans.

The capacity of lymphatic vessels to regenerate was first observed nearly a century ago.[4–6] Pronounced lymphatic growth has been detected in wound healing and inflammation, and some studies have reported an increase of lymphatic vessel density associated with certain tumors.[3,6,7] These findings did not receive much attention, mainly for two reasons. First, the credibility of the findings has been questioned, due to the lack of markers that would have allowed reliable distinction of lymphatic from blood vasculature. Second, the lymphatic system has traditionally been assigned a passive role in cancer metastasis; therefore, any significance of these findings has been questioned. Consequently, while the importance of the lymphatic system as a pathway for metastasis has been well recognized, there is very little information available about the mechanisms by which tumor cells interact with the lymphatics.

The function of lymphatic vessels in cancer remains an area of controversy. One debate revolves around the question of whether tumor lymphangiogenesis exists. Very few studies have addressed this issue altogether, and only recently has evidence been obtained for tumor lymphangiogenesis by using novel molecular markers of lymphatics. However, it remains an open question whether lymphangiogenesis is a common event during tumorigenesis and whether it is restricted only to certain types of cancer and/or tumor stages. Furthermore, the overall significance of lymphangiogenesis for tumor progression in autochthonous human tumors still needs to be examined.

Another hotly debated issue concerns the presence and biological significance of intratumoral lymphatic vessels. Some early studies reported intratumoral lymphatic vessels in certain types of cancer.[5–7] However, this has been interpreted mainly as a co-option of pre-existing lymphatic vessels by invading tumor cells, and it has been proposed that lymphatic vessels are absent from most tumors.[8–12] Recently, this view has begun to change, as more specific markers of lymphatic vessels have become available.[13] Nevertheless, the issue is far from being resolved, and the functional significance of intratumoral lymphatics is the subject of particularly vigorous debate.[14] Finally, the prevailing view has been that the lymphatic system plays a passive role in the metastatic process, although the evidence directly supporting or opposing this concept is lacking. It remains to be determined whether lymphatic vasculature has an active role in promoting tumor cell metastasis.

ROLE OF VEGF-C IN TUMOR LYMPHANGIOGENESIS AND METASTASIS

Vascular endothelial growth factor-C (VEGF-C), a novel member of the VEGF family of growth factors,[15,16] was the first lymphangiogenic factor identified. There is ample evidence for expression of VEGF-C in human tumors. VEGF-C has been shown to be expressed in breast,[17,18] colon,[19,20] lung,[18,21,22] thyroid,[23–25] gastric,[26] and squamous cell cancers[18]; in mesotheliomas[27] as well as neuroblastomas[28]; and in sarcomas and melanomas.[18] Moreover, a correlation between VEGF-C expression and rate of metastasis to lymph nodes has been found in breast,[17] colorectal,[19] gastric,[26] thyroid,[23,24] lung,[22] and prostate[29] cancers. Expression of the lymphangiogenic factor VEGF-C in tumors has suggested, for the first time, an active interaction between tumor cells and lymphatics.

The question as to the significance of VEGF-C expression for tumor progression remains unresolved. To address this question and assess the functional importance of lymphangiogenesis for cancer metastasis, we have engineered genetically fluorescent MDA-MB-435/GFP human breast cancer cells to overexpress VEGF-C.[30] Using this orthotopic breast cancer model in immunosuppressed mice, we demonstrated that VEGF-C increased peritumoral and intratumoral lymphatic vessel density. Overexpression of VEGF-C also resulted in significant enlargement of peritumoral, but not intratumoral lymphatic vessels. Whereas the overall lymphatic vessel density was fairly consistent between the tumors, considerable regional heterogeneity within individual tumors was found. This was not a result of regional differences in VEGF-C expression, and may instead reflect local variations of the tumor microenvironment, such as differences in extracellular matrix composition and/or mechanical forces. Because integration of lymphatic vessels with the interstitium is critical for lymphatic function,[3,4,31] the extracellular microenvironment is also likely to be critical for lymphangiogenesis. Importantly, VEGF-C-mediated increase of tumor lymphangiogenesis resulted in enhanced tumor metastases to regional lymph nodes and lungs. The degree of tumor metastases was highly correlated with the intratumoral lymphatic vessel density as well as with the depth of lymphatic vessel invasion into the tumors.[30]

Our findings demonstrate a causal role for lymphangiogenesis in tumor metastasis, and provide a mechanistic explanation for the reported correlation of VEGF-C expression in primary tumors with high incidence of metastases in patients. VEGF-C may promote metastasis by increasing the number of lymphatic vessels in the vicinity of tumor cells, thereby creating increased opportunities for tumor cells to leave the primary tumor site. It is also a possibility that the activation of lymphatics by VEGF-C could induce secretion of chemokines and similar factors by the lymphatic endothelium, thereby attracting tumor cells and facilitating their entry into lymphatics. Therefore, even when the tumor itself lacks lymphatic vessels, as in the VEGF-C-expressing pancreatic cancer,[32] an activation of peritumoral lymphatics could explain increased tumor metastasis.

What mechanism accounts for the increase of lung metastases observed in the VEGF-C-expressing breast cancer model? The accelerated growth of tumor cells in the lung is an unlikely answer, in that VEGF-C overexpression did not confer a growth advantage to tumor cells at the primary tumor site, lymph nodes, or *in vitro*, although this cannot be fully excluded. Furthermore, since VEGF-C did not increase angiogenesis in this tumor model, it is unlikely that increased access to blood vasculature accounted for the increase in metastases. One apparent possibility is that the increased lung metastasis resulted from an increased dissemination of tumor cells via the lymphatics. From the lymph node, tumor dissemination can occur via efferent lymphatics or through lymphatic-venous communications within the nodes, and subsequently via the blood stream.[33] The lymphatic system is optimally adapted for the entry and transport of cells, and therefore has many advantages over the blood circulation as a transport route for a metastasizing tumor cell or embolism.[3,5] The smallest lymphatic vessels are still much larger than blood capillaries, and flow velocities are orders of magnitude slower. Lymph fluid is similar to interstitial fluid and promotes cell viability. In contrast, tumor cells in the bloodstream experience serum toxicity, high shear stresses, and mechanical deformation, leading to an extremely low success rate of metastasis.[2,34] The preferential metastasis via lymphatics, due to expression of lymphangiogenic factors in tumors for example, might therefore promote survival of disseminating tumor cells and consequently increase their metastatic efficiency.

That distant metastases may increase as a consequence of increased lymphatic spread does not imply, as recently suggested,[35] that the lymphatic pathway is the exclusive pathway for metastasis in a particular tumor model. For example, in the MDA-435 breast cancer model, 100% of the mice presented with lung metastases, whereas approximately 50% of the mice had lymph node metastases at the given time point.[30] This indicates that at least a fraction of tumor metastases in the lung were blood-borne. Quantitative analysis revealed, however, that the total lung area containing metastases was greater in mice bearing VEGF-C-overexpressing tumors, suggesting a contribution of the lymphatic pathway to the overall lung tumor burden as well. While many types of tumor cells are capable of spreading through both blood and lymphatic vasculature, the relative contributions of the lymphatic and hematogenous pathways for dissemination of a particular tumor are difficult to assess. In conformity with the above concept, the fact that blocking VEGFR-3 did not reduce lung metastases in a lung tumor model[36] suggests that the spread to lungs of the given tumor is predominantly hematogenous and that VEGFR-3 signaling was not implicated.

ROLE OF VEGF-C IN TUMOR ANGIOGENESIS AND MACROPHAGE RECRUITMENT

Whereas overexpression of VEGF-C in the breast cancer model selectively induced tumor lymphangiogenesis,[30] studies of human melanoma xenotransplants revealed the induction of both lymphangiogenesis and angiogenesis by tumor-derived VEGF-C.[37] These distinct biological effects can be explained by the differential proteolytic processing of VEGF-C in the two tumor types. In breast cancer, the major VEGF-C form detected was the secreted 31-kDa protein that activates VEGFR-3 expressed mainly by the lymphatic vessels.[38–40] Accordingly, phosphorylation of VEGFR-3, but not VEGFR-2, was markedly increased in these tumors, which resulted in selective induction of lymphangiogenesis. In contrast, melanomas produced mainly the mature 21-kDa form of VEGF-C,[37] which activates VEGFR-3 and VEGFR-2 on lymphatic and blood vasculature respectively,[38] resulting in increased angiogenesis in addition to lymphangiogenesis. These results demonstrate that the biological effects of VEGF-C in tumors are critically dependent on the proteolytic processing. Processing to the 21-kDa protein was observed only *in vivo*, indicating a crucial role of host cells in regulating this process.[37]

Lymphatic vessels, serving as a pathway for the trafficking of leukocytes, are also important components of the immune system. Recent evidence suggests that, in addition to its effects on the vasculature, VEGF-C might also have a more direct impact on immune functions.[37] VEGFR-3 expression was detected on macrophages *in vitro* and *in vivo*, and VEGF-C-induced macrophage chemotaxis in a dose-dependent manner. In agreement with these results, expression of VEGFR-3 has been reported in certain hematopoietic and leukemia cells.[41] VEGF-C also increased peritumoral macrophage densities in melanoma xenografts.[37] These findings identify a novel function of VEGF-C as an immunomodulator and suggest its possible proinflammatory activities.

TUMOR LYMPHANGIOGENESIS: FACTS AND CONTROVERSIES

The prevailing belief to date has been that lymphangiogenesis does not take place in cancer and that lymphatic vessels are absent from most tumors.[11,12,14,42] Very few studies, however, have addressed these issues, and comprehensive evidence in favor of or against this widely accepted view is lacking. Identification of molecular markers of lymphatic vessels has now made it possible to re-examine the established views. What is the evidence for absence or presence of lymphatics in tumors? Using the novel molecular markers, lymphatic channels have so far been observed in eight different ex-

TABLE 1. Evidence for presence of intratumoral lymphatics

Tumor Type	Marker	Reference
Experimental tumors		
Melanoma/CAM	Prox-1	Papoutsi et al.[47] (2000)
Pancreatic cancer/CAM	Prox-1	Papoutsi et al.[48] (2001)
MDA-MB-435 breast cancer control and VEGF-C	LYVE-1/VEGFR-3	Skobe et al.[30] (2001)
MeWo melanoma/VEGF-C-transfected	LYVE-1/VEGFR-3	Skobe et al.[37] (2001)
293EBNA/VEGF-D	LYVE-1	Stacker et al.[49] (2001)
MCF-7/VEGF-C breast cancer	LYVE-1/VEGFR-3	Karpanen et al.[50] (2001)
MCF-7/VEGF-C breast cancer	LYVE-1/VEGFR-3	Mattila et al.[57] (2002)
A431/SCC	LYVE-1/Prox-1	Wigle et al.[52] (2002)
Human tumors		
Breast cancer	Podoplanin	Schoppmann et al.[53] (2001)
Head and neck SCC	LYVE-1	Beasley et al.[54] (2002)

TABLE 2. Evidence for absence of intratumoral lymphatics

Tumor Type	Marker	Reference
Experimental tumors		
Rip1Tag2 x RipVEGF-C transgenic mice	LYVE-1/VEGFR-3	Mandriota et al.[32] (2001)
MeWo melanoma	LYVE-1/VEGFR-3	Skobe et al.[37] (2001)
Human tumors		
Cervical cancer	Podoplanin	Birner et al.[55] (2001)
Ovarian cancer	Podoplanin	Birner et al.[56] (2000)
Melanoma	CD31+ / PAL-E-	De Waal et al.[57] (1997)
Uveal melanoma	CD31+ / PAL-E-	Clarijs et al.[58] (2001)
Liver cancer	LYVE-1/Prox-1	Carreira et al.[59] (2001)

perimental tumor models, in autochthonous human breast cancers, and in head and neck squamous cell carcinomas (TABLE 1). No evidence for intratumoral lymphatics has been found in human melanomas, cervical, ovarian, and liver carcinomas, or in the experimental models of pancreatic cancer and melanoma (TABLE 2). Future studies are required to determine whether intratumoral lymphatics are restricted only to certain types of cancer and whether their presence in tumors has any prognostic significance.

Most recently, the significance of intratumoral lymphatics for tumor dissemination has been called into question.[14,43] Lymphatic vessels that were identified in tumors using molecular markers of lymphatics could not be detected using lymphangiography, a technique that involves injection of labeled macromolecules into the interstitium for uptake into lymphatic capillaries. Based on these results, it has been concluded that tumors contained no functional lymphatics and that tumor cells can therefore not utilize intratumoral lymphatics for transport to the lymph nodes.[43] However, functional impairment of lymphatic vessels with respect to fluid and macromolecular transport is not the only possible explanation for the absence of detectable perfusion of lymphatics in tumors. Fluid and macromolecules travel through tissues according to hydrostatic and oncotic pressure gradients, following the pathways of least resistance to transport.[44] Elastic fibers, for example, represent a low resistance path for interstitial transport of fluid and are thus regarded to as pre-lymphatic pathways.[45] In normal tissues, extracellular matrix fibers are ideally arranged for directing fluid into the vessels.[45] In tumors, however, the extracellular matrix composition and organization are commonly altered and it is plausible that the fluid channels in tumor stroma are not directing fluid into the tumor lymphatics in such an organized manner. Furthermore, elevated interstitial fluid pressure has been reported in tumors,[46] resulting in steep hydrostatic pressure gradients at the tumor edge that may force fluid primarily out of the tumor and not laterally into tumor lymphatics. While the impairment of lymphatic ability to take up fluid could explain the absence of lymphatic vessel perfusion in tumors, it is yet to be determined which of the above hypotheses provides the correct answer.

Importantly, whereas the functional state of tumor lymphatic vessels with respect to the efficient uptake of fluids and macromolecules is of great importance for overall tumor physiology and drug delivery,[46] it may not be crucial for tumor dissemination. Because of the lack of detectable accumulation of an interstitially injected tracer in tumor lymphatics, it has been interpreted that these can not be utilized by tumor cells for migration to lymph nodes.[14,43] Such a conclusion is based on the assumption that the transport of fluid and cells in tissues and their uptake into the lymphatics is governed by the same principles. This is unlikely, as cell migration in tissues is a tightly controlled process involving a defined set of cell interactions with their microenvironment, such as responsiveness to soluble factors, attachment to specific components of the extracellular matrix, and localized proteolysis. In contrast, the major forces controlling uptake of fluid and macromolecules into lymphatics are pressure gradients in tissue.[44] Whether these forces have any effect on cell transport into the lymphatics remains an open question. Hence, the formation of an intratumoral lymphatic network, whether fully functional in fluid uptake or not, may still promote metastatic tumor spread by creating increased opportunities for metastatic tumor cells to leave the primary tumor site. Moreover, lymphatic vasculature may become activated in tumors and increase tumor cell propensity to metastasize. For example, activation of lymphatics by VEGF-C or VEGF-D could promote production of chemoattractants by lymphatic endothelial cells and thereby facilitate tumor cell entry into the lymphatics.

In conclusion, despite the intense discussion revolving around the presence and significance of intratumoral lymphatics, these might not be the most relevant questions to be answered. It is evident that tumor cells can utilize peritumoral lymphatics to spread; therefore, intratumoral lymphatics should be regarded as an additional pathway rather than a necessity for metastasis. A more fundamental issue to be addressed is the mechanism by which tumor cells enter lymphatics, regardless of their location. Our recent data indicate an active role of lymphatic endothelium in tumor metastasis, the underlying mechanisms of which are a subject of ongoing investigation.

REFERENCES

1. ZETTER, B.R. 1993. Adhesion molecules in tumor metastasis. Sem. Cancer Res. **4:** 219–229.
2. LIOTTA, L.A., P.S. STEEG & W.G. STETLER-STEVENSON. 1991. Cancer metastasis and angiogenesis: an imbalance of positive and negative regulation. Cell **64:** 327–336.
3. SWARTZ, M.A. & M. SKOBE. 2001. Lymphatic function, lymphangiogenesis, and cancer metastasis. Microsc. Res. Tech. **55:** 92–99.
4. PULLINGER, B.D. & H.W.FLOREY. 1935. Some observations on the structure and functions of lymphatics: their behavior in local edema. Br. J. Exp. Pathol. **16:** 49.
5. WITTE, M.H., D.L. WAY, C.L.WITTE, et al. 1997. Lymphangiogenesis: mechanisms, significance and clinical implications. Experientia **79:** 65–112.
6. REICHERT, F.L. 1926. The regeneration of the lymphatics. Arch. Surg. **13:** 871–881.
7. EVANS, H.M. 1908. On the occurence of newly formed lymphatic vessels in malignant growths. Johns Hopkins Med. J. **19:** 232.
8. LEE, F.C. & R.C. TILGHMANN. 1933. Lymph vessels in rabbit carcinoma, with a note on the normal lymph vessel structure of the testis. Arch. Surg. **26:** 602–616.
9. GILCHRIST, R.K. 1950. Surgical management of advanced cancer of the breast. Arch. Surg. **61:** 913–929.
10. ZEIDMAN, I., B.E. COPELAND & S. WARREN. 1955. Experimental studies on the spread of cancer in the lymphatic system. II. Absence of a lymphatic supply in carcinomas. Cancer **8:** 123–127.
11. FOLKMAN, J. 1996. Angiogenesis and tumor growth. N. Engl. J. Med. **334:** 921.
12. LEU, A.J., D.A. BERK, A. LYMBOUSSAKI, et al. 2000. Absence of functional lymphatics within a murine sarcoma: a molecular and functional evaluation. Cancer Res **60:** 4324–4327.
13. SLEEMAN, J.P., J. KRISHNAN, V. KIRKIN, et al. 2001. Markers for the lymphatic endothelium: in search of the holy grail? Microsc. Res. Tech. **55:** 61–69.

14. JAIN, R.K. & B.T. FENTON. 2002. Intratumoral lymphatic vessels: a case of mistaken identity or malfunction? J. Natl. Cancer Inst. **94:** 417–421.
15. JOUKOV, V., K. PAJUSOLA, A. KAIPAINEN, et al. 1996. A novel vascular endothelial growth factor, VEGF-C, is a ligand for the Flt4 (VEGFR-3) and KDR (VEGFR-2) receptor tyrosine kinases. EMBO J. **15:** 290–298.
16. LEE, J., A. GRAY, J. YUAN, et al. 1996. Vascular endothelial growth factor-related protein: a ligand and specific activator of the tyrosine kinase receptor Flt4. Proc. Natl. Acad. Sci. USA **93:** 1988–1992.
17. KUREBAYASHI, J., T. OTSUKI, H. KUNISUE, et al. 1999. Expression of vascular endothelial growth factor (VEGF) family members in breast cancer. Jpn. J. Cancer Res. **90:** 977–981.
18. SALVEN, P., A. LYMBOUSSAKI, P. HEIKKILA, et al. 1998. Vascular endothelial growth factors VEGF-B and VEGF-C are expressed in human tumors. Am. J. Pathol. **153:** 103–108.
19. AKAGI, K., Y. IKEDA, M. MIYAZAKI, et al. 2000. Vascular endothelial growth factor-C (VEGF-C) expression in human colorectal cancer tissues. Br. J. Cancer **83:** 887–891.
20. ANDRE, T., L. KOTELEVETS, J.C. VAILLANT, et al. 2000. Vegf, Vegf-B, Vegf-C and their receptors KDR, FLT-1 and FLT-4 during the neoplastic progression of human colonic mucosa. Int. J. Cancer **86:** 174–181.
21. NIKI, T., S. IBA, M. TOKUNOU, et al. 2000. Expression of vascular endothelial growth factors A, B, C, and D and their relationships to lymph node status in lung adenocarcinoma. Clin. Cancer Res. **6:** 2431–2439.
22. OHTA, Y., H. NOZAWA, Y. TANAKA, et al. 2000. Increased vascular endothelial growth factor and vascular endothelial growth factor-c and decreased nm23 expression associated with microdissemination in the lymph nodes in stage I non-small cell lung cancer. J. Thorac. Cardiovasc. Surg. **119:** 804–813.
23. BUNONE, G., P. VIGNERI, L. MARIANI, et al. 1999. Expression of angiogenesis stimulators and inhibitors in human thyroid tumors and correlation with clinical pathological features. Am. J. Pathol. **155:** 1967–1976.
24. FELLMER, P.T., K. SATO, R. TANAKA, et al. 1999. Vascular endothelial growth factor-C gene expression in papillary and follicular thyroid carcinomas. Surgery **126:** 1056–1061.
25. SHUSHANOV, S., M. BRONSTEIN, J. ADELAIDE, et al. 2000. VEGF-C and VEGFR3 expression in human thyroid pathologies. Int. J. Cancer **86:** 47–52.
26. YONEMURA, Y., Y. ENDO, H. FUJITA, et al. 1999. Role of vascular endothelial growth factor C expression in the development of lymph node metastasis in gastric cancer. Clin. Cancer Res. **5:** 1823–1829.
27. OHTA, Y., V. SHRIDHAR, R.K. BRIGHT, et al. 1999. VEGF and VEGF type C play an important role in angiogenesis and lymphangiogenesis in human malignant mesothelioma tumours. Br. J. Cancer **81:** 54–61.
28. EGGERT, A., N. IKEGAKI, J. KWIATKOWSKI, et al. 2000. High-level expression of angiogenic factors is associated with advanced tumor stage in human neuroblastomas. Clin. Cancer Res. **6:** 1900–1908.
29. TSURUSAKI, T., S. KANDA, H. SAKAI, et al. 1999. Vascular endothelial growth factor-C expression in human prostatic carcinoma and its relationship to lymph node metastasis. Br. J. Cancer **80:** 309–313.
30. SKOBE, M., T. HAWIGHORST, D.G. JACKSON, et al. 2001. Induction of tumor lymphangiogenesis by VEGF-C promotes breast cancer metastasis. Nat. Med. **7:** 192–198.

31. LEAK, L.V. 1970. Electron microscopic observations on lymphatic capillaries and the structural components of the connective tissue-lymph interface. Microvasc. Res. **2:** 361–391.
32. MANDRIOTA, S.J., L. JUSSILA, M. JELTSCH, et al. 2001. Vascular endothelial growth factor-C-mediated lymphangiogenesis promotes tumour metastasis. EMBO J. **20:** 672–682.
33. FISHER, B. & E.R. FISHER. 1968. Role of the lymphatic system in dissemination of tumor. *In* Lymph and the Lymphatic System. H.S. Mayerson, Ed.: 324. Charles C. Thomas. Springfield, IL.
34. WEISS, L. & G.W. SCHMID-SCHÖNBEIN. 1989. Biomechanical interactions of cancer cells with the microvasculature during metastasis. Cell Biophys. **14:** 187–215.
35. JAIN, R.K. & T.P. PADERA. 2002. Prevention and treatment of lymphatic metastasis by antilymphangiogenic therapy. J. Natl. Cancer Inst. **94:** 785–787.
36. HE, Y., K. KOZAKI, T. KARPANEN, et al. 2002. Suppression of tumor lymphangiogenesis and lymph node metastasis by blocking vascular endothelial growth factor receptor 3 signaling. J. Natl. Cancer Inst. **94:** 819–825.
37. SKOBE, M., L.M. HAMBERG, T. HAWIGHORST, et al. 2001. Concurrent induction of lymphangiogenesis, angiogenesis, and macrophage recruitment by vascular endothelial growth factor-C in melanoma. Am. J. Pathol. **159:** 893–903.
38. JOUKOV, V., T. SORSA, V. KUMAR, et al. 1997. Proteolytic processing regulates receptor specificity and activity of VEGF-C. EMBO J. **16:** 3898–3911.
39. KAIPAINEN, A., J. KORHONEN, T. MUSTONEN, et al. 1995. Expression of the fms-like tyrosine kinase 4 gene becomes restricted to lymphatic endothelium during development. Proc. Natl. Acad. Sci. USA **92:** 3566–3570.
40. JUSSILA, L., R.VALTOLA, T.A. PARTANEN, et al. 1998. Lymphatic endothelium and Kaposi's sarcoma spindle cells detected by antibodies against the vascular endothelial growth factor receptor-3. Cancer Res. **58:** 1599–1604.
41. FIELDER, W., U. GRAEVEN, S. ERGUN, et al. 1997. Expression of FLT4 and its ligand VEGF-C in acute myeloid leukemia. Leukemia **11:** 1234–1237.
42. CARMELIET, P. & R.K. JAIN. 2000. Angiogenesis in cancer and other diseases. Nature **407:** 249–257.
43. PADERA, T.P., A. KADAMBI, E.DI TOMASO, et al. 2002. Lymphatic metastasis in the absence of functional intratumor lymphatics. Science **296:** 1883–1886.
44. SCHMID-SCHÖNBEIN, G.W. 1990. Microlymphatics and lymph flow. Physiol. Rev. **70:** 987–1028.
45. RYAN, T.J. 1989. Structure and function of lymphatics. J. Invest. Dermatol. **93:** 18S–24S.
46. JAIN, R.K. 1990. Vascular and interstitial barriers to delivery of therapeutic agents in tumors. Cancer Metast. Rev. **9:** 253–266.
47. PAPOUTSI, M., G. SIEMEISTER, K. WEINDEL, et al. 2000. Active interaction of human A375 melanoma cells with the lymphatics in vivo. Histochem. Cell Biol. **114:** 373–385.
48. PAPOUTSI, M., J.P. SLEEMAN & J. WILTING. 2001. Interaction of rat tumor cells with blood vessels and lymphatics of the avian chorioallantoic membrane. Microsc. Res. Tech. **55:** 100–107.
49. STACKER, S.A., C. CAESAR, M.E. BALDWIN, et al. 2001. VEGF-D promotes the metastatic spread of tumor cells via the lymphatics. Nat. Med. **7:** 186–191.

50. KARPANEN, T., M. EGEBLAD, M.J. KARKKAINEN, et al. 2001. Vascular endothelial growth factor C promotes tumor lymphangiogenesis and intralymphatic tumor growth. Cancer Res. **61:** 1786–1790.
51. MATTILA, M.M., J.K. RUOHOLA, T. KARPANEN, et al. 2002. VEGF-C induced lymphangiogenesis is associated with lymph node metastasis in orthotopic MCF-7 tumors. Int. J. Cancer **98:** 946–951.
52. WIGLE, J.T., N. HARVEY, M. DETMAR, et al. 2002. An essential role for Prox1 in the induction of the lymphatic endothelial cell phenotype. EMBO J. **21**(7): 1505–1513.
53. SCHOPPMANN, S.F., P. BIRNER, P. STUDER & S. BREITENEDER-GELEFF. 2001. Lymphatic microvessel density and lymphovascular invasion assessed by anti-podoplanin immunostaining in human breast cancer. Anticancer Res. **21**(4A): 2351–2355.
54. BEASLEY, N.J., R. PREVO, S. BANERJI, et al. 2002. Intratumoral lymphangiogenesis and lymph node metastasis in head and neck cancer. Cancer Res. **62**(5): 1315–1320.
55. BIRNER, P., M. SCHINDL, A. OBERMAIR, et al. 2001. Lymphatic microvessel density as a novel prognostic factor in early-stage invasive cervical cancer. Int. J. Cancer **95:** 29–33.
56. BIRNER, P., M. SCHINDL, A. OBERMAIR, et al. 2000. Lymphatic microvessel density in epithelial ovarian cancer: its impact on prognosis. Anticancer Res. **20:** 2981–2985.
57. DE WAAL, R.M., M.C. VAN ALTENA, H. ERHARD, et al. 1997. Lack of lyphangiogenesis in human primary cutaneous melanoma. Consequences for the mechanism of lymphatic dissemination. Am. J. Pathol. **150**(6): 1951–1957.
58. CLARIJS, R., L. SCHALKWIJK, D.J. RUITER & R.M. DE WAAL, et al. 2001. Lack of lymphangiogenesis despite coexpression of VEGF-C and its receptor Flt-4 in uveal melanoma. Invest. Ophthalmol. Visual Sci. **42**(7): 1422–1428.
59. CARREIRA, C.M., S.M. NASSER, E. DI TOMASO, et al. 2001. LYVE-1 is not restricted to the lymph vessels: expression in normal liver blood sinusoids and down-regulation in human liver cancer and cirrhosis. Cancer Res. **61:** 8079–8084.

The Pathogenesis of Filarial Lymphedema

Is it the Worm or Is It the Host?

PATRICK J. LAMMIE,[a] KAREN T. CUENCO,[b] AND
GEORGE A. PUNKOSDY[a,c]

[a]*Division of Parasitic Diseases, Centers for Disease Control and Prevention, Atlanta, Georgia 30341, USA*

[b]*Department of Human Genetics, University of Pittsburgh School of Public Health, Pittsburgh, Pennsylvania, USA*

[c]*Department of Cell Biology, University of Georgia, Athens, Georgia, USA*

> ABSTRACT: Our understanding of the pathogenesis of filarial lymphedema, although evolving, is still limited. Recurrent bacterial infections play a major role in the progression of lymphedema to elephantiasis, but the host and parasite factors that trigger disease development are not known. Field studies in Haiti show that lymphedema and host responses to parasite antigens cluster in families, consistent with the hypothesis that host genes influence lymphedema susceptibility. The recent recognition that filarial parasites harbor the endosymbiotic bacteria, *Wolbachia*, also raises questions about the potential contribution of the inflammatory response to *Wolbachia* antigens to lymphedema development. In this review, we discuss potential risk factors for lymphedema and try to integrate these in a model of pathogenesis.
>
> KEYWORDS: Lymphatic filariasis; *Wuchereria bancrofti*; lymphedema

BACKGROUND

Lymphatic filariasis is a significant public health problem throughout tropical areas of the world, affecting more than 100 million people, millions of whom have elephantiasis or hydrocele. Filarial infection is mosquito-transmitted, but the efforts to control transmission that are based exclusively on mosquito control have had limited success. New global efforts to eliminate transmission of lymphatic filariasis have been facilitated by development of

---

Address for correspondence: Patrick J. Lammie, Ph.D., Division of Parasitic Diseases/F13, Centers for Disease Control and Prevention, 4770 Buford Highway, Atlanta, GA 30341. Voice: 770-488-4054; fax: 770-488-4108.
PJL1@cdc.gov

rapid diagnostic tests (used to identify foci of infection) and by the demonstration of the effectiveness of mass treatment strategies with drugs that target microfilariae, the transmission stages of the parasite found in the blood.

Lymphedema and elephantiasis are major sequelae of lymphatic filariasis affecting up to 2% of the population in areas of intense transmission. The majority of affected persons are women, who experience social isolation, decreased productivity, and debilitating recurrent attacks of adenolymphangitis. Risk factors for this disease process are not well understood. Adult stages of the parasites develop and reside in the lymphatic vessels of humans; however, persons with clinical evidence of lymphedema frequently have no evidence of active infection, raising a number of questions about the nature of disease development. Several distinct, but not mutually exclusive, models of disease pathogenesis have been proposed.[1–3] In this review, we will discuss risk factors for lymphedema and try to integrate these in a model of disease development.

ESTABLISHMENT OF A STABLE HOST–PARASITE RELATIONSHIP

The establishment of the parasite within the lymphatic vessels of the host is the key event in the cascade that ultimately leads to pathology, yet, in the case of filarial parasites, we know little about the interactions that govern this process, either from the perspective of the parasite or that of the host. Filarial parasites are quite host specific. Although strains of *Brugia malayi* are known that can infect a variety of host species, *Wuchereria bancrofti*, the species responsible for 90% of lymphatic filariasis in humans, is difficult to establish in animals, even in primate models of infection. What governs this host specificity? Is the localization of developing filarial larvae within lymphatic vessels an entirely passive process? Does *W. bancrofti* have receptors for ligands produced by lymphatic endothelium or other host cells that provide physiologic cues or developmental signals?[4]

Our knowledge of the location of *W. bancrofti* in humans has been aided greatly by Dr. Gerusa Dreyer's pioneering work with ultrasound in Brazil.[5] With the use of ultrasound as a diagnostic tool, it was possible to address questions about the distribution of adult worms in humans and their impact on the lymphatics. These studies established that lymphatic vessels associated with the spermatic cord provide a preferred location for *W. bancrofti* in adult men; however, worms have been much more difficult to localize in adult women.[6–8]

Ultrasonographic studies also suggest that lymphatic dilatation is a consistent feature of worm infection, but lymphangiectasia also is found in vessels where no signs of worm activity are present.[2] Interestingly, when ultrasound is used, filarial worms in young boys are found in the inguinal and other peripheral lymph nodes, but not in the spermatic cord.[9,10] It is not clear to what extent these results reflect the true location of the worms, rather than an artifact related to the ease of finding evidence of worm movement in dilated lymphatic vessels by ultrasound. Nonetheless, these observations imply either that localization of worms or development of lymphangiectasia is influenced by changes in host hormone levels or other physiologic cues affected by puberty.

The role of the host immune response in influencing the establishment of initial filarial infection is also not clearly understood.[11] Paradoxically, immunologically naive persons seem to be more resistant to *W. bancrofti* infection than persons who have grown up in an endemic area where they have been exposed to infection over a long period of time. Patent infections develop much more commonly among persons born in an endemic area than among visitors or relocated persons. More than 30 years ago, Beaver proposed "the possibility of tolerance induced by individual and direct contact with the filaria during the fetal and neonatal periods of development" as an explanation for this difference.[12] These observations imply that host immune responses may represent permissive as well as non-permissive factors for parasite development (FIG. 1).

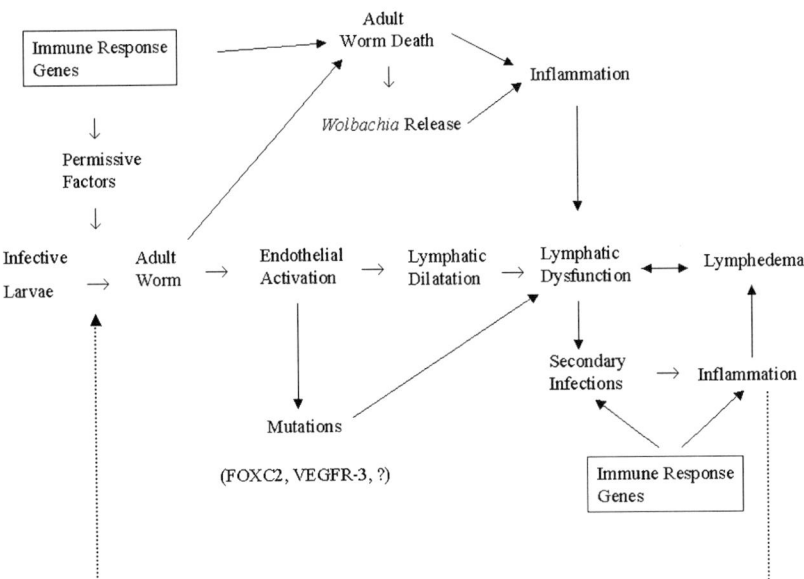

FIGURE 1. A schematic of factors that may contribute to the development of filarial lymphedema. Evidence for each of these is discussed in the text. Independent of events involved in triggering of lymphedema, recurrent bacterial infections represent a common risk factor for exacerbation of lymphedema and progression to elephantiasis.

Ravindran has suggested that Th1- and Th2-biased immune responses provide different developmental cues for larval stages of filarial parasites and hypothesized that Th1-skewed responses provide an environment that is permissive for larval development.[13] Since the absence of infection is associated with Th1 responses,[14,15] and L3 larvae induce Th2 responses *in vivo* very rapidly following infection,[16] we would argue the reverse; that is, we propose that Th2 responses are permissive for larval development. *In utero* exposure to filarial antigen may lead to downregulation of the expression of Th1-associated responses, thus facilitating establishment of patent infections.[17,18] It is important to bear in mind that Th1 and Th2 responses may not be affecting the parasite directly, but may be associated with other host factors that influence parasite development.

THE ROLE OF HOST IMMUNITY IN PATHOGENESIS

The observation that the prevalence of microfilaremia is significantly lower among persons with lymphedema than among persons without overt disease suggested the possibility of a relationship between clearance of infection, expression of immunity, and disease development.[19,20] Consistent with this idea, a number of early studies showed that antifilarial responses, especially cell-mediated immune responses, are elevated in patients with lymphedema, compared with microfilaremic persons.[21-24] This work was done prior to the development of antigen assays and before it was recognized that antifilarial immune responses were profoundly influenced by antigen status.[14,15] To investigate the relationship of antifilarial immunity to disease more directly, we compared antifilarial immune responses of Haitian women with lymphedema and antigen-negative women without evidence of lymphatic disease.[25] Lymphedema patients had significantly higher levels of antifilarial-specific IgG1, IgG2, and IgG3 than antigen-negative women with no evidence of lymphatic disease. These results demonstrated that high levels of antifilarial immunity are associated with the presence of lymphedema and are consistent with a potential role for antifilarial immune reactivity in disease development or progression. We hypothesize that parasite antigen stimulation of T cells is associated with increased local inflammatory responses that promote lymphatic damage and contribute to pathogenesis.

THE ROLE OF SECONDARY INFECTIONS AND ACUTE ATTACKS

Episodes of adenolymphangitis (ADL) are a significant cause of morbidity and have been suggested to be a risk factor for the development and progression of lymphedema and elephantiasis, but the causes of ADL have not been clearly defined. Based on clinical criteria, two distinct syndromes of ADL have been characterized by Dreyer and colleagues.[26] A syndrome of descending or retrograde ADL is related to the death of the adult worm and is frequently associated with a discrete nodule or cord-like lesion in the lymphatic vessel. In contrast, a diffuse or reticular pattern of ADL is associated with loss of skin integrity or entry lesions and is related to secondary bacterial infections. The proportion of ADL due to these two syndromes in different settings is unknown, although evidence collected in Brazil suggests that bacterial ADL may be as much as 30-fold more common than filarial ADL.

Acute attacks of bacterial ADL last 2–5 days, usually affect the lower limbs, and are associated with transient exacerbations of lymphedema.[27–31] Bacterial ADL is more common among persons with existing lymphedema than among persons with no outward manifestations of lymphatic damage, consistent with the idea that recurrent infections with bacterial agents may be a key determinant of disease progression.[3,32] Bacteria can be isolated from the blood and lymph of persons experiencing acute attacks, and acute ADL attacks are prevented by improving skin hygiene and by treating entry lesions with topical antibiotics and antifungal creams.[27,31,33] These observations support the hypothesis that bacterial infections contribute to progression of "filarial" disease.

The causative agents of bacterial ADL have not been fully characterized. Both pathogenic streptococcal species and normal skin flora have been implicated as causes of ADL. A number of investigators have analyzed acute and convalescent sera collected from ADL cases to determine whether changes in antibody responses to specific agents are consistently associated with ADL attacks.[34,35] In our studies, we observed significant changes in antibody to one or more streptococcal antigens in 30% of paired acute and convalescent serum samples.[36] Further studies are needed to determine what organisms are responsible for triggering ADL and whether exposure to filarial larvae plays a direct or indirect role in this process.

THE ROLE OF HOST GENETICS

Although the establishment of adult worms is thought to lead, in virtually all cases, to the development of lymphangiectasia, most persons with long-term infections are clinically asymptomatic.[3,7] Only a small proportion of endemic residents develop lymphedema, which may progress to elephantiasis. As noted above, persons with lymphedema also differ from other persons in endemic communities in terms of the prevalence of filarial infection and the expression of antifilarial immunity. These observations are consistent with the hypothesis that a small proportion of exposed persons may be genetically predisposed to disease development.

To test this hypothesis, we collected pedigree information from persons attending a lymphedema treatment program at a hospital in Haiti who were designated as affected probands. Pedigrees were analyzed to determine whether the families observed to have multiple lymphedema cases had a higher prevalence of lymphedema than expected when lymphedema prevalence estimates and family size were considered. Forty-four of 173 pedigrees studied had more than one lymphedema case.[37] The number of families demonstrating excess disease was significantly different from what would have been expected using population estimates of lymphedema prevalence ($P < 0.05$), and lymphedema of the leg was excessive in 15 of 44 sampled families with multiple lymphedema cases. The presence of excess disease in a proportion of the families supports the general hypothesis that lymphedema of the leg aggregates within families.

In follow-up studies, we assessed familial clustering of antibody responses to filarial and bacterial antigens. Unaffected controls and their households were selected from the same neighborhood as those of affected probands. Antibody responses to filarial antigen and streptolysin O (SLO) were assessed for the family members of probands and controls by enzyme-linked immunosorbant assay ($n = 748$). Using multiple linear regression, we found that antifilarial IgG2 antibody was significantly correlated with infection status ($P < 0.001$), with lymphedema status ($P < 0.001$), and with the kinship coefficient ($P = 0.026$), a measure of the genetic relatedness to the proband.[38] Although secondary bacterial infections are a risk factor for acute adenolymphangitis (ADL) and lymphedema development, we found no correlation between kinship and the antibody response to the two streptococcal antigens we tested. These results suggest that both antifilarial immunity and lymphedema cluster in families and that immune responses to filarial antigens may be associated with lymphedema development.

Recent exciting results have provided a better understanding of the genetic basis of primary and hereditary lymphedema syndromes and suggest a new focus for studies of filarial lymphedema. Patients with primary lymphedema, an autosomal dominant condition, develop lymphedema early in life and have a similar predisposition to recurrent bacterial infection as patients with filarial lymphedema. Development of primary lymphedema is associated with point mutations in the gene for the vascular endothelial growth factor receptor 3 (VEGFR-3).[39] Mice heterozygous for a VEGFR-3 mutation lack subcutaneous lymphatic vessels and develop a lymphedema-like syndrome that can be reversed by over-expression of VEGF.[40] VEGFR-3 mutations in humans with primary lymphedema cluster in the tyrosine kinase domain and interfere with VEGF-induced cellular activation.[39] As all patients characterized to date are heterozygous for the mutation, it is thought that the mutant receptor competes with the wild-type receptor, leading to altered endothelial cell function and disrupted lymphangiogenesis.

Hereditary lymphedema represents a complex group of autosomal dominant syndromes associated with different ages of onset and clinical findings. These different syndromes were shown to be associated with mutations in the forkhead transcription factor, FOXC2.[41] All lymphedema-associated FOXC2 mutations described to date result in the production of a truncated protein. Interestingly, identical mutations have been linked to different clinical phenotypes, suggesting that we do not have a good understanding of the relationship between the development of lymphedema and FOXC2 expression.

Filarial lymphedema, like hereditary lymphedema, is pleiotropic in terms of age of onset and speed of progression. Our pedigree analyses show that clustering of lymphedema occurs in a proportion of affected families, especially those with both affected male and female family members. Furthermore, elevated antifilarial immune responses also cluster in families of persons with lymphedema. Based on these observations, we hypothesize that filarial lymphedema represents the interaction between filarial infection and two sets of genetic risk factors: (1) immune response genes associated with heightened inflammatory responses and lymphatic damage following filarial infection; and (2) loss-of-function mutations in the VEGFR-3, FOXC2, or related genes associated with impaired lymphangiogenesis (FIG. 1). The high-risk families that we have identified in Haiti provide an important opportunity to test these hypotheses.

THE ROLE OF *WOLBACHIA* RELEASE FOLLOWING WORM DEATH

Despite the dissociation between lymphedema and current infection (that is, microfilaremia and antigenemia), it seems likely that lymphedema patients were infected by adult worms at some point prior to lymphedema development. As a trigger of inflammatory reactions, death of the adult worm may represent a critical point in development of lymphatic pathology. It is not clear how this event is related to the transition from Th2-immune responses, which are associated with active infection, to the Th1-like antifilarial responses associated with lymphedema.[14] Interestingly, filarial worms are known to harbor *Wolbachia,* rickettsia-like organisms that live symbiotically within the worms. This raises the question whether *Wolbachia* bacteria or antigens contribute to the inflammatory cascade seen in lymphedema patients and provide the missing link between the patient's heightened responses to worm and bacterial antigens (FIG. 1).

Wolbachia are transmitted vertically in filariae and, thus, are found in all stages of the parasite. Following chemotherapy, *Wolbachia* DNA can be detected in the serum and *Wolbachia* lipopolysaccharide (LPS) is thought to cause systemic adverse reactions.[42,43] We have been interested in the possibility that the host response to *Wolbachia* may play a role in immunopathogenesis, as originally hypothesized by Kozek.[44] We have demonstrated that *Brugia malayi*-infected rhesus monkeys mount antibody responses to *Wolbachia* surface protein (WSP) that are temporally associated with worm death and lymphedema development.[45] In addition, patients with lymphedema are also significantly more likely to have serum antibodies to WSP than individuals without clinical evidence of lymphatic disease ($P = 0.0005$).[46] These results suggest that additional investigations of *Wolbachia* may be important in furthering our understanding of the development of lymphedema.

AN ATTEMPT AT SYNTHESIS

Previous discussions of the pathogenesis of filarial lymphedema have tended to focus on single pathways of disease development. Based on the discussions above and evidence of clustering of lymphedema in some family pedigrees, we now consider it more likely that multiple pathways are involved in the disease process (FIG. 1). Resolving this complexity will require a great deal of additional effort. To date, published clinical studies of filarial lymphedema have been focused on clinical staging, but not on age of onset, the level of immune responsiveness, frequency of acute attacks, or other characteristics of the disease that may be useful for clarifying pathogenic mechanisms. Study populations such as that in Haiti provide an excellent opportunity to conduct a more detailed clinical characterization of patients with lymphedema from high- and low-risk families and to screen these families for genes that may be linked to lymphedema development. In addition, it is still difficult to understand why persons with lymphedema typically remain antigen negative and thus, presumably, uninfected in the face of ongoing exposure to mosquitoes carrying infective larvae. Do inflammatory responses associated with recurrent bacterial infection suppress larval development as we suggest in FIGURE 1? Addressing this issue may shed light on mechanisms of protective immunity in lymphatic filariasis. Finally, independent of the initial events involved in the triggering of lymphedema development, heightened susceptibility to recurrent bacterial infection appears to represent a common pathway for disease progression. Thus, clinical management of lymphedema should focus on prevention of the acute attacks that exacerbate disease.

PUBLIC HEALTH CONTEXT

Because of the recent development of effective single-dose drug treatment, global elimination of lymphatic filariasis is now considered possible. The success of campaigns to eliminate lymphatic filariasis will depend, at least in part, on our ability to motivate populations in endemic areas to make a long-term commitment to the effort. Developing and maintaining this support may be problematic, because side-effects associated with microfilaria clearance following treatment may diminish compliance. In addition, patients with lymphedema, the group bearing the greatest burden of disease in the community, are typically microfilaria-negative and derive no direct benefit from chemotherapy. It is important to develop new strategies to treat lymphedema so that these efforts can be incorporated into the program to eliminate this parasite. These strategies should be based on a solid understanding of the pathogenesis of lymphedema.

ACKNOWLEDGMENTS

Financial support for this work was generously provided, in part, through graduate student awards from Emory University and the University of Georgia and CDC Emerging Infectious Diseases funds.

REFERENCES

1. OTTESEN, E.A. 1984. Immunological aspects of lymphatic filariasis and onchocerciasis in man. Trans. R. Soc. Trop. Med. Hyg. **78** (Suppl.): 9–18.
2. FREEDMAN, D.O. 1998. Immune dynamics in the pathogenesis of human lymphatic filariasis. Parasitol. Today **14**: 229–234.
3. DREYER, G., J. NOROES, J. FIGUEREDO-SILVA & W.F. PIESSENS. 2001. Pathogenesis of lymphatic disease in Bancroftian filariasis: a clinical perspective. Parasitol. Today **16**: 544–548.
4. RAJAN, T.V. 1998. A hypothesis for the tissue specificity of nematode parasites. Exp. Parasitol. **89**: 140–142.
5. AMARAL, F., G. DREYER, J. FIGUEREDO-SILVA, J. NOROES, et al. 1994. Live adult worms detected by ultrasonography in human Bancroftian filariasis. Am. J. Trop. Med. Hyg. **50**: 753–757.
6. NOROES, J., D. ADDISS, F. AMARAL, A. COUTINHO, et al. 1996. Occurrence of living adult *Wuchereria bancrofti* in the scrotal area of men with microfilaremia. Trans. R. Soc. Trop. Med. Hyg. **90**: 55–56.
7. NOROES, J., D. ADDISS, A. SANTOS, Z. MEDEIROS, et al. 1996. Ultrasonographic evidence of abnormal lymphatic vessels in young men with adult *Wuchereria bancrofti* infection in the scrotal area. J. Urol. **156**: 409–412.
8. DREYER, G., A.C. BRANDAO, F. AMARAL, et al. 1996. Detection by ultrasound of living adult *Wuchereria bancrofti* in the female breast. Mem. Inst. Oswaldo Cruz **91**: 95–96.
9. FURNESS, B.W., J.K. TRAYNER, J.M. BRISSAU, et al. 2002. Ultrasonic examination of Haitian children with lymphatic filariasis. Manuscript in preparation.

10. DREYER, G., J. NOROES, D. ADDISS, et al. 1999. Bancroftian filariasis in a paediatric population: an ultrasonographic study. Trans. R. Soc. Trop. Med. Hyg. **93:** 633–636.
11. MAIZELS, R.M., J.E. ALLEN & M. YAZDANBAKHSH. 1999. Immunology of lymphatic filariasis: current controversies. *In* Lymphatic Filariasis. T.B. Nutman, Ed.: 217–243. Imperial College Press. London.
12. BEAVER, P.C. 1970. Filariasis without microfilaremia. Am. J. Trop. Med. Hyg. **19:** 181–189.
13. RAVINDRAN, B. 2001. Are inflammation and immunological hyperactivity needed for filarial parasite development. Trends Parasitol. **17:** 70–73.
14. DIMOCK, K.A., M.L. EBERHARD & P.J. LAMMIE. 1996. Th1-like antifilarial immune responses predominate in antigen-negative persons. Infect. Immun. **64:** 2962–2967.
15. DEALMEIDA, A.B., M.C. MAIA E SILVA, M.A. MACIEL & D.O. FREEDMAN. 1996. The presence or absence of active infection, not clinical status, is most closely associated with cytokine responses in lymphatic filariasis. J. Inf. Dis.**173:** 1453–1459.
16. OSBORNE, J. & E. DEVANEY. 1998. The L3 of *Brugia* induces a Th2-polarized response following activation of an IL-4-producing CD4-CD8-$\alpha\beta$ T cell population. Int. Immunol. **10:** 1583–1590.
17. HIGHTOWER, A.W., P.J. LAMMIE & M.L. EBERHARD. 1993. Maternal filarial infection — a persistent risk factor for microfilaremia in offspring. Parasitol. Today **9:** 418–421.
18. STEEL, C., A. GUINEA, J.S. MCCARTHY & E.A. OTTESEN. 1994. Long-term effect of prenatal exposure to maternal microfilaremia on immune responsiveness to filarial parasite antigens. Lancet **343:** 890–893.
19. LAMMIE, P.J., D.G. ADDISS, G. LEONARD, et al. 1993. Heterogeneity of filarial-specific immune responsiveness among patients with lymphatic obstruction. J. Infect. Dis. **167:** 1178–1183.
20. ADDISS, D.G., K.A. DIMOCK, M.L. EBERHARD & P.J. LAMMIE. 1995. Clinical, parasitologic, and immunologic observations of patients with hydrocele and elephantiasis in an area with endemic lymphatic filariasis. J. Infect. Dis. **171:** 755–758.
21. OTTESEN, E.A., P.F. WELLER & L. HECK. 1977. Specific cellular immune unresponsiveness in human filariasis. Immunology **33:** 413–421.
22. PIESSENS, W.F., P.B. MCGREEVY, P.W. PIESSENS, et al. 1980. Immune response in human infections with *Brugia malayi*: specific cellular unresponsiveness to filarial antigens. J. Clin. Invest. **65:** 172–177.
23. OTTESEN, E.A. 1992. Infection and disease in lymphatic filariasis: an immunological perspective. Parasitology **104:** S71–S79.
24. MAIZELS, R.M., E. SARTONO, A. KURNIAWAN, et al. 1995. T cell activation and the balance of antibody isotypes in human lymphatic filariasis. Parasitol. Today **11:** 50–56.
25. BAIRD, J.B., P.J. LAMMIE, J. LOUIS CHARLES, et al. 2002. Reactivity to bacterial, fungal, and parasite antigens in patients with lymphedema and elephantiasis. Am. J. Trop. Med. Hyg. **66:** 163–169.
26. DREYER, G., Z. MEDEIROS, M.J. NETTO, et al. 1999. Acute attacks in the extremities of persons living in an area endemic for Bancroftian filariasis: differentiation of two syndromes. Trans. R. Soc. Trop. Med. Hyg. **93:** 413–417.

27. SHENOY, R.K., K. SANDHYA, T.K. SUMA & V. KUMARASWAMI. 1995. A preliminary study of filariasis-related acute adenolymphangitis with special reference to precipitating factors and treatment modalities. S.E. Asian J. Trop. Med. Pub. Health **26**: 301–305.
28. GYAPONG, J.O., M. GYAPONG & S. ADJEI. 1996. The epidemiology of acute adenolymphangitis due to lymphatic filariasis in northern Ghana. Am. J. Trop. Med. Hyg. **54**: 591–595.
29. RAMAIAH, K.D., K. RAMU, K.N.V. KUMAR & H. GUYATT. 1996. Epidemiology of acute filarial episodes caused by *Wuchereria bancrofti* infection in two rural villages in Tamil Nadu, south India. Trans. R. Soc. Trop. Med. Hyg. **90**: 639–643.
30. ALEXANDER, N.D.E., R.T. PERRY, Z. DIMBER, et al. 1999. Acute disease episodes in a *Wuchereria bancrofti*-endemic area of Papua New Guinea. Am. J. Trop. Med. Hyg. **64**: 319–324.
31. OLSZEWSKI, W.L., S. JAMAL, G. MANOKARAN, et al. 1999. Bacteriological studies of blood, tissue fluid, lymph and lymph nodes in patients with acute dermatolymphangioadenitis (DLA) in course of 'filarial'lymphedema. Acta Tropica **73**: 217–224.
32. DREYER, G. & W.F. PIESSENS. 1999. Worms and microorganisms can cause disease in residents of filariasis-endemic areas. *In* Lymphatic Filariasis. T.B. Nutman, Ed.: 239–256. Imperial College Press. London.
33. SEIM, A.A., G. DREYER & D.G. ADDISS. 1999. Controlling morbidity and interrupting transmission: twin pillars of lymphatic filariasis elimination. Rev. Soc. Bras. Med. Trop. **32**: 325–328.
34. VINCENT, A.L., C.A. URENA ROJAS, E.M. AYOUB, et al. 1998. Filariasis and erisipela in Santo Domingo. J. Parasitol. **84**: 557–561.
35. ESTERRE, P., C. PLICHART, M.O. HUIN-BLONDEY & L. NGUYEN. 2000. Role of streptococcal infection in the acute pathology of lymphatic filariasis. Parasite **2**: 91–94.
36. GOODMAN, D. P.J. LAMMIE, J. BAIRD, et al. 2002. Antibacterial immune reactivity in lymphedema and adenolymphangitis. Manuscript in preparation.
37. T.CUENCO, K. 2001. Familial clustering of filarial lymphedema in a Haitian population. Ph.D. dissertation. Emory University. Atlanta, GA.
38. YANG, N., K. T.CUENCO & P.J. LAMMIE. 2002. Familial studies of lymphedema and immune responsiveness to filarial antigens. Manuscript in preparation.
39. KARKKAINEN, M.J., R.E. FERRELL, E.C. LAWRENCE, et al. 2000. Missense mutations interfere with VEGFR-3 signalling in primary lymphedema. Nat. Gen. **25**: 153–159.
40. KARKKAINEN, M.J., A. SAARISTO, L. JUSSILA, et al. 2001. A model for gene therapy of human hereditary lymphedema. Proc. Natl. Acad. Sci. USA **98**: 12677–12682.
41. FINEGOLD, D.N., M.A. KIMAK, E.C. LAWRENCE, et al. 2001. Truncating mutation in FOXC2 cause multiple lymphedema syndromes. Hum. Mol. Gen. **10**: 1185–1189.
42. CROSS, H.F., M. HAARBRINK, G. EGERTON, et al. 2001. Severe reactions to filarial chemotherapy and release of *Wolbachia* endosymbionts into blood. Lancet **358**: 1873–1875.
43. KEISER, P.B., S.M. REYNOLDS, K. AWADZI, et al. 2002. Bacterial endosymbionts of *Onchocerca volvulus* in the pathogenesis of post-treatment reactions. J. Infect. Dis. **185**: 805–811.

44. KOZEK, W.J. 1977. Transovarially transmitted intracellular microorganisms in adult and larval stages of *Brugia malayi*. J. Parasitol. **63:** 992–1000.
45. PUNKOSDY, G.A., V.A. DENNIS, B. LASATER, *et al.* 2001. Detection of serum IgG antibodies specific for *Wolbachia* surface protein in rhesus monkeys infected with *Brugia malayi*. J. Inf. Dis.**184:** 385–389.
46. PUNKOSDY, G.A., D.G. ADDISS & P.J. LAMMIE. 2002. Characterization of antibody responses to *Wolbachia* surface protein (WSP) among Haitians with lymphatic filariasis. Manuscript in preparation.

The Molecular Control of Adipogenesis, with Special Reference to Lymphatic Pathology

EVAN D. ROSEN

Beth Israel Deaconess Medical Center, Boston, Massachusetts 02215, USA

ABSTRACT: Adipogenesis is the process by which mature fat cells are formed from pre-adipocytes. Adipogenesis has come under increasing scrutiny not only because the availability of reliable *in vitro* models makes it an attractive choice for developmental studies, but also because adipocytes are increasingly recognized as major players in a variety of physiological and pathophysiological states, such as obesity and type 2 diabetes. Adipocytes develop from mesenchymal stem cell precursors that are characterized by multipotency. Under the influence of various cues, these cells become committed to the adipocyte lineage. Further hormonal stimulation recruits these pre-adipocytes to accumulate lipid, express fat-specific markers, and become sensitive to the metabolic effects of insulin. A complex transcriptional cascade regulates this process, involving several distinct classes of transcription factor. In particular, the role of the nuclear hormone receptor PPARγ will be discussed, along with bZip family members C/EBPα, C/EBPβ, and C/EBPδ. The relationship of adipose depots to the lymphatic system will also be discussed.

KEYWORDS: adipogenesis; PPARγ; C/EBP; lymphedema; lymph nodes

WHY STUDY ADIPOGENESIS?

In the not-so-distant past, few tissues were felt to be as unworthy of study as adipose tissue. Fat has been routinely ignored and discarded by anatomists and physiologists, while the general public has invested adipocytes with more emotional baggage than any other non-neoplastic cell type. In the last decade or so, the tide has begun to turn in favor of adipose tissue as a subject of scientific inquiry. The reasons for this are threefold. First, the development of immortal preadipocyte cell lines by Green and colleagues in the 1970s has

Address for correspondence: Evan D. Rosen, M.D., Ph.D., Beth Israel Deaconess Medical Center, 99 Brookline Ave., Boston, MA 02215. Voice: 617-667-3221; fax: 617-667-2927.
erosen@caregroup.harvard.edu

made tractable the study of adipocyte development and physiology.[1,2] There are few other developmental processes that can be reliably studied in a nearly synchronous population of cells *in vitro*. Compared to many other cellular lineages, fat cell differentiation *in vitro* is rather authentic, recapitulating most of the key features of adipogenesis *in vivo*. This includes morphological changes associated with lipid accumulation, cessation of cell growth, expression of many lipogenic enzymes, and establishment of sensitivity to most or all of the key hormones that impact on this cell type. The second reason for the recent surge in popularity of adipocytes is the recognition that these cells are far more than passive depots for energy storage. Fat is now known to secrete a wide variety of proteins that influence energy homeostasis (such as leptin and adiponectin),[3–5] coagulation (such as PAI-1),[6,7] blood pressure (such as angiotensinogen),[8] and immune function (such as TNF-α, IL-6, adipsin).[9,10] These findings firmly establish fat as an endocrine organ to be reckoned with. Finally, the dramatic rise in the incidence of obesity and type 2 diabetes worldwide has also focused attention on the biology of the adipocyte.[11]

HOW DO ADIPOCYTES DEVELOP?

Adipose development is unusual in that it occurs *de novo* in multiple, anatomically distinct sites. Generally speaking, most adipose tissues form at sites rich in loose connective tissues, such as the subcutaneous layers between the muscle and dermis. However, fat deposits form in many locations, such as around the heart, kidneys, and other internal organs. Older studies have described subcutaneous fat development in some morphological detail. In the absence of a clear molecular marker for pre-adipocytes, the earliest event associated with adipogenesis is a proliferating network of capillaries in an otherwise undistinguished region of loose connective tissue. These "primitive organs," as they have been called, develop into *bona fide* adipose tissue.[12] These observations also point out a potentially important relationship between blood vessel development and adipogenesis. It is not clear, however, whether adipogenesis can induce angiogenesis, vice versa, or both.

Adipocytes develop from multipotent mesenchymal stem cells that can also give rise to muscle, bone, or cartilage[13,14] (FIG. 1). In the developing fat pad, these cells become committed to the adipocytic lineage under the influence of "cues" that remain undiscovered. These cues might be hormonal interactions, or the result of cell–cell or cell–matrix interactions, for example. This process, called determination, results in a cell with fibroblastic morphology called a pre-adipocyte. Unfortunately, there are no known expression markers that absolutely identify a cell as a pre-adipocyte. Because of the complexity inherent in these systems, almost all work on adipogenesis has utilized predetermined clonal cell lines, such as 3T3-L1 and 3T3-F442A. Upon reaching confluence, exposure to a regimen including dexamethasone, methylisobutylxanthine (a phosphodiesterase inhibitor), and insulin (DMI) leads to differentiation of these cells.

Adipogenesis occurs in both the prenatal and postnatal states in humans; in rodents, most fat cell development occurs postnatally. While older literature suggested that people are born with all the adipocytes they will possess in life, there is now convincing evidence that adipogenesis occurs throughout the lifetime of an organism. This adipogenesis occurs both as a consequence of normal cell turnover, and as a consequence of the requirement for additional fat mass that arises with significant calorie storage and weight gain.[15] Indeed, while fat cell size can vary with the amount of lipid stored, there is a physical limit to how large these cells can become. On the other hand, humans and other animals will continue to gain fat as long as energy intake exceeds nutritional requirements, demonstrating a theoretical requirement for *de novo* differentiation of adipocytes. More convincing evidence has come from studies of rats fed a high-calorie diet, where [^{3}H]thymidine incorporation into new fat cells occurs throughout adulthood.[16] Whether this is true for humans is unknown, but it is clear that pre-adipocytes purified from the fat pads of even elderly people can be induced to differentiate *in vitro*.[17–19] This observation argues for a role for both adipocyte hypertrophy *and* hyperplasia in human obesity.

HORMONAL AND ENVIRONMENTAL ASPECTS OF ADIPOGENESIS

The literature is rife with papers identifying extracellular and intracellular signals that control pre-adipocyte growth and terminal differentiation. The first agent noted to affect this process was insulin, the classic hormone of the fed state. Insulin increases the percentage of cells that differentiate and also increases the amount of lipid that each fat cell contains.[20] Recent data have shown that neutralization of insulin *in vivo* stimulates the apoptosis of fat cells,[21] dramatically confirming the importance of this hormone in adipocyte biology. Glucocorticoids have also been used for many years to induce optimal differentiation of cultured pre-adipose cell lines and primary adipocytes. It is not as clear, however, if glucocorticoids are required for efficient adipogenesis *in vivo*; since patients with Cushing's syndrome (elevated circulating glucocorticoids, usually from a pituitary or adrenal source) show visceral obesity but wasting of subcutaneous fat.[22] The effects of other hormones are also not completely understood. Growth hormone, for example, can clearly induce adipogenesis in a variety of cultured pre-adipose cell lines, but primary pre-adipocytes do not show this effect.[23–25] In fact, differentiation of these cells appears to be inhibited by growth hormone. This is consistent with the observation that humans with growth hormone deficiency have normal fat stores, and can, in fact, be obese. Thyroid hormone,[26] retinoic acid,[27] and various prostaglandins[28] are some of the wide variety of other hormones that have been shown to affect adipogenesis *in vitro*, but for which there is scant evidence to support such a role *in vivo*.

A variety of cytokines have been found to suppress fat cell differentiation. TNF-α, IL-1, and many other pro-inflammatory cytokines have this effect on most cultured pre-adipocyte lines, and in fact can "de-differentiate" already mature fat cells.[29,30] Indeed, the suppression of adipocyte lipoprotein lipase was used to purify "cachectin," which was then found to be identical to TNF-α[31] The precise molecular basis for these cytokine effects is still not known, though suppression of expression of several key transcriptional regulators of adipogenesis has been observed. In addition, several growth factors are rather potent inhibitors of adipogenesis, including epidermal growth factor (EGF), platelet-derived growth factor (PDGF), and fibroblast growth factor (FGF).[32] Much of this inhibition is likely to be through activation of mitogen-activated protein (MAP) kinases (see below).

As might be expected from the wide array of hormones, cytokines, and growth factors that can affect adipogenesis, several different signal transduction pathways have been implicated in the process as well. There has been quite a bit of interest in the MAP kinase family of intracellular protein kinases. As just described, many growth factors that exert their influence through tyrosine kinase receptors inhibit fat cell differentiation. It is now clear that much of this inhibition occurs through activation of the "classic" MAP kinases, Erk 1 and 2. These kinases directly phosphorylate PPARγ (see subsequent text), and inhibit its adipogenic activity.[33] The JNK MAP kinase can also phosphorylate PPARγ at this inhibitory site.[34] In contrast, p38 MAP kinase has been associated with an activation of adipogenesis. Chemical p38 inhibitors and a dominant negative allele of p38 decrease fat cell differentiation, and overexpression of constitutively active MKK6, an activator of p38, enhance the process.[35,36] The relevant targets of p38 are unknown at this time, but speculation centers on C/EBPβ, an early actor in the adipogenic transcriptional cascade. Another kinase, Akt (also known as PKB) has been shown to be a downstream effector of certain of the metabolic effects of insulin. Interestingly, expression of a constitutively active form of Akt causes spontaneous differentiation of 3T3-L1 cells.[37] The experimental manipulation of cAMP, various G proteins, and protein kinase C isoforms have also been shown to affect adipogenesis *in vitro*.[38] A more thorough knowledge of the critical extracellular signals involved in fat cell development would better enable us to place these findings in their appropriate context.

Another environmental condition that affects adipogenesis is hypoxia, which inhibits the process. This effect seems to be mediated through

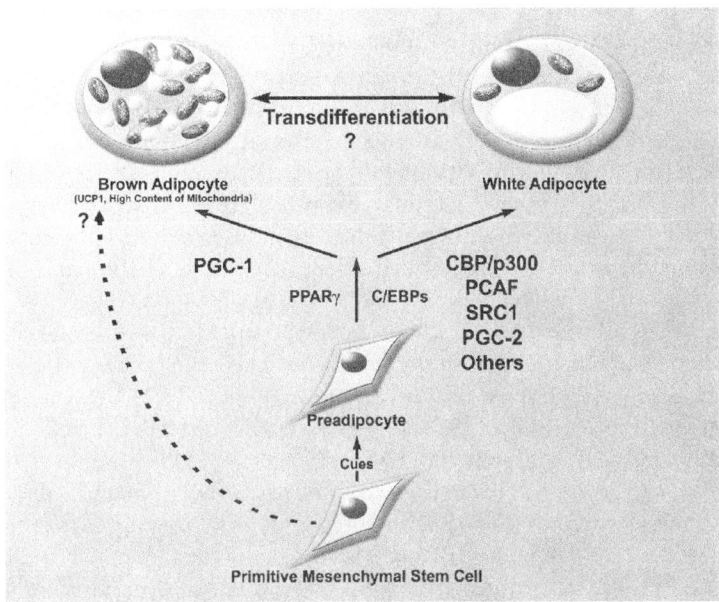

FIGURE 1. Adipogenesis proceeds from mesenchymal stem cells in two major steps. First, the multipotent stem cells become committed to the adipocytic lineage in a process called determination. Next, the committed pre-adipocytes fully differentiate through the concerted actions of a host of transcriptional regulators (see text for details). It is not known whether brown and white adipocytes use similar pathways for adipogenesis. There is some evidence that these two cell types can "*trans*-differentiate" into one another. Alternatively, brown adipocytes may develop along an independent pathway.

hypoxia-inducible factor 1α (HIF-1α), which represses transcriptional induction of PPARγ via an intermediate factor, called DEC1/Stra13.[39]

TRANSCRIPTIONAL CONTROL OF ADIPOSE DIFFERENTIATION

Studies in the pre-adipose cell lines 3T3-L1 and 3T3-F442A have yielded enormous insight into the transcriptional cascade that operates during adipose conversion (FIG. 2). Several groups have published data on a handful of transcription factors that are intimately involved in the transition from pre-adipocyte to adipocyte. We know, for example, that members of the CCAAT/ enhancer binding protein (C/EBP) family are important in this process. C/EBPβ and C/EBPδ are expressed early in adipogenesis *in vitro*, reaching maximum levels within the first 2 days after differentiation is induced.[40, 41]

These factors in turn induce C/EBPα and the nuclear hormone receptor peroxisome proliferator-activated receptor γ (PPARγ), whose levels peak after a few days of differentiation and remain elevated throughout the life of the adipocyte. The ectopic addition of either C/EBPα or C/EBPβ is sufficient to drive the complete program of adipogenesis in pre-adipocyte cell culture models; C/EBPδ is not sufficient per se, but does accelerate differentiation induced by C/EBPβ[42] *In vivo,* the total loss of C/EBPα is associated with neonatal lethality resulting from hepatic and other defects[43]; until recently; this has hampered a careful study of the effect of loss of C/EBPα on postnatal events such as fat cell development. New studies have now been performed with alternative strategies that rescue the hepatic phenotype; a strong case has been made that C/EBPα is required for the normal development of white adipose tissue (WAT), but not for brown adipose tissue (BAT).[44,45] Interestingly, combined loss of C/EBPβ and C/EBPδ has the converse effect—modest reductions in WAT with more dramatic defects in BAT.[46]

A role for PPARγ in adipogenesis was first suggested by analysis of the aP2 (the adipocyte-specific fatty-acid-binding protein) 5′-flanking region. Elements were identified in this promoter region that were shown to be both necessary and sufficient to direct gene expression from a minimal promoter specifically to fat cells in culture and in transgenic mice.[47,48] Cloning of the cognate *trans*-acting factor identified it as a heterodimer consisting of two nuclear receptors, PPARγ and RXRα (retinoid X receptor-α)[49] PPARγ exists as two isoforms created by alternative splicing at the 5′ end of a single gene; PPARγ2 contains 30 amino acids more at the N-terminus than does PPARγ1. While many tissues express a low level of the PPARγ1 protein, fat expresses high levels of PPARγ2 (and also PPARγ1).

As a member of the nuclear hormone receptor superfamily, PPARγ is activated through the binding of ligands to a distinct carboxyl-terminal ligand-binding domain. It is now appreciated that the synthetic thiazolidinedione (TZD) class of anti-diabetic drugs function as ligands for PPARγ[50] These molecules, which have affinity (K_D) in the 50–700 nM range, were originally developed through medicinal chemistry approaches, with no insight into their mechanism of action. While these compounds are very effective at promoting adipogenesis in culture and *in vivo*, it is not known whether this effect is related to the anti-diabetic actions of these drugs. The endogenous PPARγ ligand is not known. Certain fatty acids can bind PPARγ with low affinity, but most investigators feel that these compounds are not present at physiological concentrations in nuclei *in vivo*. An unusual prostanoid, 15-deoxy-$\Delta^{12,14}$-PGJ2 (prostaglandin J2), has also been identified as a ligand for PPARγ[51,52] but it is not known if this compound even exists in nature. This prostanoid has also been shown to activate other, unrelated signal transduction pathways in cells (such as NF-κB), making it difficult to ascribe its effects as belonging to PPARγ.[53]

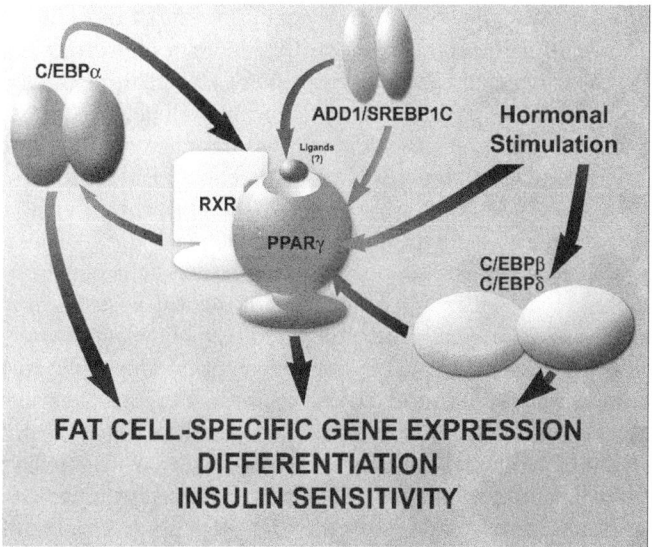

FIGURE 2. Transcriptional cascade in adipogenesis. Under the influence of unknown cues, C/EBPβ and C/EBPδ rise early during the conversion of pre-adipocytes to adipocytes. These factors induce the expression of C/EBPα and PPARγ, each of which maintains the expression of the other. Together these factors (and ADD1/SREBP1c) induce terminal gene expression and the final differentiated phenotype.

PPARγ plays a crucial role in the function of many fat-cell selective genes. PPARγ binding is absolutely required for the function of the fat-selective enhancers for the *FABP4* (aP2) and *PEPCK* genes in cultured fat cells.[54] This analysis of PEPCK has recently been extended *in vivo*, where the expression of the gene in fat was shown to be dependent on PPARγ binding, while expression in other tissues was not.[55] In addition to directly binding to and regulating fat cell-selective genes, PPARγ plays a dominant role in the regulation of differentiation per se. This was shown first in "gain of function" experiments that expressed PPARγ in non-adipogenic, fibroblastic cells (NIH-3T3) using retroviral vectors.[56] While ligands selective for PPARγ were not known at this time, application of the pan-PPAR ligand ETYA (5,8,11,14-eicosatetraynoic acid) activated PPARγ and caused a powerful differentiation response. The use of high-affinity TZD ligands has greatly improved these experiments. PPARγ-mediated differentiation included lipid accumulation and the expression of many endogenous genes characteristic of this cell type. The ability of PPARγ to promote adipogenesis is not limited to fibroblastic cells. Myoblastic cell lines can also be converted to adipocytes, particularly when the cells co-express both C/EBPα and PPAR.[57]

Loss-of-function studies with PPARγ have been more difficult to perform, because mice with targeted ablation of PPARγ die *in utero* at embryonic day 9.5–10.[58,59] This occurs because of a failure of placentation, precluding useful analysis of the role played by this factor in adipogenesis, most of which occurs postnatally in rodents. Our group has circumvented this problem by using chimeric lineage analysis to demonstrate that PPAR is required for adipogenesis *in vivo*.[60] Another group reported the same results using tetraploid rescue of the placental defect in PPARγ knockout mice.[59] These findings were extended by demonstrating that PPARγ null ES cells or primary embryonic fibroblasts were incapable of undergoing adipogenesis *in vitro*.[58,60]

The synthesis of these and other studies has led to the following model: C/EBPβ and C/EBPδ are activated by unknown cues in the committed pre-adipocyte to induce PPARγ and C/EBPα. These two factors then induce each other's expression in a positive feedback loop that maintains the differentiated state. We and others have shown that cells lacking C/EBPα can in fact be differentiated to adipocytes as long as PPARγ is added ectopically (although they lack normal insulin sensitivity).[61,62] This raises the question of whether C/EBPα and PPARγ represent redundant pathways in adipogenesis or whether the role of C/EBPα is primarily directed at inducing and maintaining PPARγ levels. We now have data from PPARγ null fibroblasts that shows that C/EBPα *cannot* rescue the adipogenic phenotype in these cells.[63] The role of C/EBPα in adipogenesis may in fact be more ancillary than had been previously thought; at least one major function of C/EBPα is to induce and maintain appropriate levels of PPARγ

In addition to this cross-regulation, it is clear that these proteins can act synergistically to activate differentiation and differentiation-linked gene expression. The molecular basis for this synergy is not known, but is worth noting that many fat cell genes have binding sites for both C/EBP and PPARγ/RXR. Possibilities for this synergy include regulation of co-activators involved in adipogenesis, and/or the possibility that C/EBPα is involved in regulatory genes that produce PPARγ ligands.

Another transcription factor that promotes adipogenesis is adipocyte determination and differentiation factor-1 (ADD1), also known as sterol regulatory element binding protein-1c (SREBP1c). ADD1/SREBP1c is induced in differentiating 3T3-L1 cells at around the same time as PPARγ and C/EBPα.[64] A constitutively active form of ADD1/SREBP1c enhances 3T3-L1 adipogenesis, while a dominant negative form represses this process.[65] There is evidence that ADD1/SREBP1c activates the PPARγ promoter,[66] and may also contribute to the generation of a PPARγ ligand.[67] Beyond these effects, ADD1/SREBP1c mediates some of the transcriptional effects of insulin in adipose tissue, particularly on lipogenic genes such as fatty acid synthase, acetyl-CoA carboxylase, and stearoyl-CoA desaturase.[68]

There have been reports of several other transcription factors that are linked to adipogenesis. Though none of these have been shown to affect adipogenesis as profoundly as PPARγ and the C/EBPs, it is quite possible that some of these will be important modulators of fat cell development and function. PPARδ in particular has been studied in this light, with conflicting results. Some workers have shown that NIH-3T3 fibroblasts stably expressing PPARδ are unable to undergo differentiation,[69] while others have shown that they can as long as a cAMP-inducing agent is added to the differentiation cocktail.[70] Similarly, claims have been made that some pre-adipocyte lines (such as Ob1771 and 3T3-F442A) can be induced to differentiate more completely if PPARδ is forcibly expressed or if a PPARδ ligand like 2-bromopalmitate is added.[71,72] Another group showed that 3T3-L1 preadipocytes, on the other hand, were only modestly affected by the addition of a more specific PPARδ ligand.[70] Loss-of-function studies with PPARδ have not shed as much light as might be hoped. Two different groups have generated PPARδ null mice; most PPARδ –/– embryos die mid-gestation, but a few survive to term and show reduced adiposity.[73, 74] Of note, mice with PPARδ targeted only in adipocytes show no decrease in adiposity, indicating that an extra-adipose site must be involved in the decreased body mass seen in the global knockout.[73] Other factors have also been discovered, including the forkhead factor foxc2 which is expressed in an adipose-selective manner and which promotes adipocyte-specific (especially BAT-specific) gene expression in a transgenic model.[75]

Finally, a few transcription factors have been shown to inhibit adipogenesis. These include GATA2 and GATA3, which are expressed in pre-adipocytes.[76] These factors work at least in part by repressing the PPARγ2 promoter. Reductions in GATA2 and GATA3 levels are required for adipogenesis to proceed *in vitro*. DFosB is a naturally occurring variant of FosB that appears to repress adipogenesis in marrow stromal cells while promoting the formation of osteoblasts.[77]

WHAT IS THE RELATIONSHIP OF ADIPOGENESIS TO LYMPHATIC PATHOLOGY?

Adipose tissue is found in discrete depots scattered through the body. Some of these depots are more metabolically active than others, and tend to expand or shrink in proportion to nutritional status. Other depots, such as the structural fat found in the heel and fingerpads, are quite metabolically inert, and barely change in size even in the leanest animals. Another place where fat can always be found, regardless of nutritional status, is around lymph nodes. These perinodal adipocytes typically do not respond to fasting, but are fully capable of activating lipolysis when stimulated by local lymph nodes or co-cultured lymphoid cells.[78,79] There are interesting studies to demonstrate that adipocytes that are physically juxtaposed to lymphatic tissue are exquisitely sensitive to the lipolytic effects of lymphoid-derived cytokines like TNF-α and interleukin-6, while other adipocytes from distal sites within the same depot do not respond to these stimuli to the same magnitude.[80] There also appear to be qualitative differences in the fatty acid composition of resident triglycerides in perinodal adipocytes relative to fat cells elsewhere.[81] These observations have led to the hypothesis that adipose tissue forms an important energy depot for local lymphoid metabolism, and that these stores are important enough that they are spared by more mundane stimuli, such as fasting. Consistent with this notion is the fact that chronic inflammatory states such as Crohn's disease can lead to redistribution of fat.[82]

Another pathological state involving the lymphatic system is lymphedema, characterized by the accumulation of interstitial lymph in a defined region of the body, often as a result of interruption of lymphatic drainage. Chronic lymphedema leads to a variety of pathological changes in the affected region, including overgrowth of connective tissue and adipose tissue.[83] This exuberant growth of adipose tissue has led some to question whether lymph possesses an adipogenic activity, although there has been remarkably little work done to address this possibility. One study showed that cultured rabbit pre-adipocytes could be induced to differentiate more completely if lymph was included in the medium.[84] The stimulatory effect could be fully replicated by adding chylomicrons to the medium in place of whole lymph. On the one hand, this observation might suggest that the phenomenon is created by pre-existing, small adipocytes that take up and incorporate the triglycerides in the chylomicrons. The effect would thus be on lipogenesis rather than on adipogenesis per se. One should not, however, discount the possibility that a *bona fide* adipogenic stimulus is present in the chylomicron fraction of lymph. An example of this might be a lipophilic ligand for PPARγ, which could be enriched in the very low-density lipoprotein (VLDL).

CONCLUSIONS

Despite its complexity, adipogenesis has proven to be a developmental system tractable to study, because of the availability of cultured cell lines that faithfully mimic fat cell development *in vivo*. Such studies have allowed the identification of several critical transcription factors that promote the differentiated state. These factors include PPARγ, ADD1/SREBP1c, and members of the C/EBP family. Despite these advances, many aspects of adipogenesis remain mysterious, including the steps that lead to commitment of the multipotent mesenchymal stem cell to the adipocytic lineage, as well as the control of terminal gene expression in mature fat cells. Additionally, several observations indicate the presence of cross-talk between adipocytes and the lymphatic system. Unraveling these mysteries will certainly lead to insights that could have benefits in conditions ranging from obesity and type 2 diabetes to lymphedema.

REFERENCES

1. GREEN, H. & O. KEHINDE. 1975. An established preadipose cell line and its differentiation in culture. II. Factors affecting the adipose conversion. Cell **5:** 19–27.
2. GREEN, H. & O. KEHINDE. 1976. Spontaneous heritable changes leading to increased adipose conversion in 3T3 cells. Cell **7:** 105–113.
3. SCHERER, P.E., S. WILLIAMS, M. FOGLIANO, et al. 1995. A novel serum protein similar to C1q, produced exclusively in adipocytes. J. Biol. Chem. **270:** 26746–22749.
4. HU, E., P. LIANG & B.M. SPIEGELMAN. 1996. AdipoQ is a novel adipose-specific gene dysregulated in obesity. J. Biol. Chem. **271:** 10697–10703.
5. ZHANG, Y., R. PROENCA, M. MAFFEI, et al. 1994. Positional cloning of the mouse obese gene and its human homologue. Nature **372:** 425–432.
6. LUNDGREN, C.H., S.L. BROWN, T.K. NORDT, et al. 1996. Elaboration of type-1 plasminogen activator inhibitor from adipocytes: a potential pathogenetic link between obesity and cardiovascular disease. Circulation **93:** 106–110.
7. SAMAD, F., K. YAMAMOTO & D.J. LOSKUTOFF. 1996. Distribution and regulation of plasminogen activator inhibitor-1 in murine adipose tissue in vivo: induction by tumor necrosis factor-alpha and lipopolysaccharide. J. Clin. Invest. **97:** 37–46.
8. SAYE, J.A., L.A. CASSIS, T.W. STURGILL, et al. 1989. Angiotensinogen gene expression in 3T3-L1 cells. Am. J. Physiol. **256:** C448–C451.
9. COOK, K.S., H.Y. MIN, D. JOHNSON, et al. 1987. Adipsin: a circulating serine protease homolog secreted by adipose tissue and sciatic nerve. Science **237:** 402–405.
10. HOTAMISLIGIL, G.S., N.S. SHARGILL & B.M. SPIEGELMAN. 1993. Adipose expression of tumor necrosis factor-alpha: direct role in obesity-linked insulin resistance. Science **259:** 87–91.
11. MOKDAD, A.H., B.A. BOWMAN, E.S. FORD, et al. 2001. The continuing epidemics of obesity and diabetes in the United States. J. Am. Med. Assoc. **286:** 1195–1200.
12. WASSERMAN, F. 1926. The fat organs of man: development, structure, and systematic place of the so-called adipose tissue. Z. Zellforsch. Mikroskop. Anat. Abt. Histochem. **3:** 325.
13. TAYLOR, S.M. & P.A. JONES. 1979. Multiple new phenotypes induced in 10T1/2 and 3T3 cells treated with 5-azacytidine. Cell **17:** 771–779.
14. PITTENGER, M.F., A.M. MACKAY, S.C. BECK, et al. 1999. Multilineage potential of adult human mesenchymal stem cells. Science **284:** 143–147.
15. PRINS, J.B. & S. O'RAHILLY. 1997. Regulation of adipose cell number in man. Clin. Sci. (London) **92:** 3–11.

16. MILLER, W.H., JR., I.M. FAUST & J. HIRSCH. 1984. Demonstration of de novo production of adipocytes in adult rats by biochemical and radioautographic techniques. J. Lipid Res. **25:** 336–347.
17. KIRKLAND, J.L., C.H. HOLLENBERG & W.S. GILLON. 1990. Age, anatomic site, and the replication and differentiation of adipocyte precursors. Am. J. Physiol. **258:** C206–C210.
18. ENTENMANN, G. & H. HAUNER. 1996. Relationship between replication and differentiation in cultured human adipocyte precursor cells. Am. J. Physiol. **270:** C1011–C1106.
19. DESLEX, S., R. NEGREL, C. VANNIER, et al. 1987. Differentiation of human adipocyte precursors in a chemically defined serum-free medium. Int. J. Obes. **11:** 19–27.
20. GIRARD, J., D. PERDEREAU, F. FOUFELLE, et al. 1994. Regulation of lipogenic enzyme gene expression by nutrients and hormones. FASEB J. **8:** 36–42.
21. KIESS, W. & B. GALLAHER. 1998. Hormonal control of programmed cell death/apoptosis. Eur. J. Endocrinol. **138:** 482–491.
22. NEWELL-PRICE, J., P. TRAINER, M. BESSER, et al. 1998. The diagnosis and differential diagnosis of Cushing's syndrome and pseudo-Cushing's states. Endocr. Rev. **19:** 647–672.
23. CATALIOTO, R.M., D. GAILLARD, G. AILHAUD, et al. 1992. Terminal differentiation of mouse preadipocyte cells: the mitogenic-adipogenic role of growth hormone is mediated by the protein kinase C signalling pathway. Growth Factors **6:** 255–264.
24. HAUSMAN, G.J. & R.J. MARTIN. 1989. The influence of human growth hormone on preadipocyte development in serum-supplemented and serum-free cultures of stromal-vascular cells from pig adipose tissue. Domest. Anim. Endocrinol. **6:** 331–337.
25. WABITSCH, M., H. HAUNER, E. HEINZE, et al. 1995. The role of growth hormone/insulin-like growth factors in adipocyte differentiation. Metabolism **44:** 45–49.
26. GHARBI-CHIHI, J., P. GRIMALDI, J. TORRESANI, et al. 1981. Triiodothyronine and adipose conversion of OB17 preadipocytes: binding to high-affinity sites and effects on fatty acid synthetizing and esterifying enzymes. J. Recept. Res. **2:** 153–173.
27. SAFONOVA, I., C. DARIMONT, E.Z. AMRI, et al. 1994. Retinoids are positive effectors of adipose cell differentiation. Mol. Cell. Endocrinol. **104:** 201–211.
28. REGINATO, M.J., S.L. KRAKOW, S.T. BAILEY, et al. 1998. Prostaglandins promote and block adipogenesis through opposing effects on peroxisome proliferator-activated receptor gamma. J. Biol. Chem. **273:** 1855–1858.
29. PETRUSCHKE, T. & H. HAUNER. 1993. Tumor necrosis factor-alpha prevents the differentiation of human adipocyte precursor cells and causes delipidation of newly developed fat cells. J. Clin. Endocrinol. Metab. **76:** 742–747.
30. OHSUMI, J., S. SAKAKIBARA, J. YAMAGUCHI, et al. 1994. Troglitazone prevents the inhibitory effects of inflammatory cytokines on insulin-induced adipocyte differentiation in 3T3-L1 cells. Endocrinology **135:** 2279–2282.
31. FRIED, S.K. & R. ZECHNER. 1989. Cachectin/tumor necrosis factor decreases human adipose tissue lipoprotein lipase mRNA levels, synthesis, and activity. J. Lipid Res. **30:** 1917–1923.
32. HAUNER, H., K. ROHRIG & T. PETRUSCHKE. 1995. Effects of epidermal growth factor (EGF), platelet-derived growth factor (PDGF) and fibroblast growth

factor (FGF) on human adipocyte development and function. Eur. J. Clin. Invest. **25:** 90–96.
33. Hu, E., J.B. Kim, P. Sarraf, *et al.* 1996. Inhibition of adipogenesis through MAP kinase-mediated phosphorylation of PPAR-gamma. Science **274:** 2100–2103.
34. Camp, H.S., S.R. Tafuri & T. Leff. 1999. c-Jun N-terminal kinase phosphorylates peroxisome proliferator-activated receptor-gamma1 and negatively regulates its transcriptional activity. Endocrinology **140:** 392–397.
35. Engelman, J.A., A.H. Berg, R.Y. Lewis, *et al.* 1999. Constitutively active mitogen-activated protein kinase kinase 6 (MKK6) or salicylate induces spontaneous 3T3-L1 adipogenesis. J. Biol. Chem. **274:** 35630–35638.
36. Engelman, J.A., M.P. Lisanti & P.E. Scherer. 1998. Specific inhibitors of p38 mitogen-activated protein kinase block 3T3-L1 adipogenesis. J. Biol. Chem. **273:** 32111–32120.
37. Magun, R., B.M. Burgering, P.J. Coffer, *et al.* 1996. Expression of a constitutively activated form of protein kinase B (c-Akt) in 3T3-L1 preadipose cells causes spontaneous differentiation. Endocrinology **137:** 3590–3593.
38. Rosen, E.D. & B.M. Spiegelman. 2000. Molecular regulation of adipogenesis. Annu. Rev. Cell Dev. Biol. **16:** 145–171.
39. Yun, Z., H.L. Maecker, R.S. Johnson, *et al.* 2002. Inhibition of PPAR-gamma2 gene expression by the HIF-1-regulated gene DEC1/Stra13: a mechanism for regulation of adipogenesis by hypoxia. Dev. Cell **2:** 331–341.
40. Yeh, W.C., Z. Cao, M. Classon, *et al.* 1995. Cascade regulation of terminal adipocyte differentiation by three members of the C/EBP family of leucine zipper proteins. Genes Dev. **9:** 168–181.
41. Cao, Z., R.M. Umek & S.L. McKnight. 1991. Regulated expression of three C/EBP isoforms during adipose conversion of 3T3-L1 cells. Genes Dev. **5:** 1538–1552.
42. Wu, Z., Y. Xie, N.L. Bucher, *et al.* 1995. Conditional ectopic expression of C/EBP beta in NIH-3T3 cells induces PPAR-gamma and stimulates adipogenesis. Genes Dev. **9:** 2350–2363.
43. Wang, N.D., M.J. Finegold, A. Bradley, *et al.* 1995. Impaired energy homeostasis in C/EBP-alpha knockout mice. Science **269:** 1108–1112.
44. Chen, S.S., J.F.Chen, P.F. Johnson, *et al.* 2000. C/EBP-beta, when expressed from the C/EBP-alpha gene locus, can functionally replace C/EBP-alpha in liver but not in adipose tissue. Mol. Cell. Biol. **20:** 7292–7299.
45. Linhart, H.G., K. Ishimura-Oka, F. DeMayo, *et al.* 2001. C/EBP-alpha is required for differentiation of white, but not brown, adipose tissue. Proc. Natl. Acad. Sci. USA **98:** 12532–12537.
46. Tanaka, T., N. Yoshida, T. Kishimoto, *et al.* 1997. Defective adipocyte differentiation in mice lacking the C/EBP-beta and/or C/EBP-delta gene. EMBO J. **16:** 7432–7443.
47. Graves, R.A., P. Tontonoz, K.A. Platt, *et al.* 1992. Identification of a fat cell enhancer: analysis of requirements for adipose tissue-specific gene expression. J. Cell. Biochem. **49:** 219–224.
48. Graves, R.A., P. Tontonoz, S.R. Ross, *et al.* 1991. Identification of a potent adipocyte-specific enhancer: involvement of an NF-1-like factor. Genes Dev. **5:** 428–437.
49. Tontonoz, P., E. Hu, R.A. Graves, *et al.* 1994. mPPAR-gamma 2: tissue-specific regulator of an adipocyte enhancer. Genes Dev. **8:** 1224–1234.

50. LEHMANN, J.M., L.B. MOORE, T.A. SMITH-OLIVER, et al. 1995. An antidiabetic thiazolidinedione is a high-affinity ligand for peroxisome proliferator-activated receptor gamma (PPAR-gamma). J. Biol. Chem. **270:** 12953–12956.
51. FORMAN, B.M., P. TONTONOZ, J. CHEN, et al. 1995. 15-Deoxy-Delta-12,14-prostaglandin J2 is a ligand for the adipocyte determination factor PPAR-gamma. Cell **83:** 803–812.
52. KLIEWER, S.A., J.M. LENHARD, T.M. WILLSON, et al. 1995. A prostaglandin J2 metabolite binds peroxisome proliferator-activated receptor gamma and promotes adipocyte differentiation. Cell **83:** 813–819.
53. CASTRILLO, A., M.J. DIAZ-GUERRA, S. HORTELANO, et al. 2000. Inhibition of I-kappa-B kinase and I-kappa-B phosphorylation by 15-deoxy-Delta(12,14)-prostaglandin J(2) in activated murine macrophages. Mol. Cell. Biol. **20:** 1692–1698.
54. TONTONOZ, P., E.HU, J. DEVINE, et al. 1995. PPAR-gamma2 regulates adipose expression of the phosphoenolpyruvate carboxykinase gene. Mol. Cell. Biol. **15:** 351–357.
55. DEVINE, J.H., D.W. EUBANK, D.E. CLOUTHIER, et al. 1999. Adipose expression of the phosphoenolpyruvate carboxykinase promoter requires peroxisome proliferator-activated receptor gamma and 9-cis-retinoic acid receptor binding to an adipocyte-specific enhancer in vivo. J. Biol. Chem. **274:** 13604–13612.
56. TONTONOZ, P., E. HU & B.M. SPIEGELMAN. 1994. Stimulation of adipogenesis in fibroblasts by PPAR-gamma2, a lipid-activated transcription factor. Cell **79:** 1147–1156.
57. HU, E., P. TONTONOZ & B.M. SPIEGELMAN. 1995. Transdifferentiation of myoblasts by the adipogenic transcription factors PPAR-gamma and C/EBP-alpha. Proc. Natl. Acad. Sci. USA **92:** 9856–9860.
58. KUBOTA, N., Y. TERAUCHI, H. MIKI, et al. 1999. PPAR-gamma mediates high-fat diet-induced adipocyte hypertrophy and insulin resistance. Mol. Cell **4:** 597–609.
59. BARAK, Y., M.C. NELSON, E.S. ONG, et al. 1999. PPAR-gamma is required for placental, cardiac, and adipose tissue development. Mol. Cell **4:** 585–595.
60. ROSEN, E.D., P. SARRAF, A.E. TROY, et al. 1999. PPAR-gamma is required for the differentiation of adipose tissue in vivo and in vitro. Mol. Cell **4:** 611–617.
61. WU, Z., E.D. ROSEN, R. BRUN, et al. 1999. Cross-regulation of C/EBP-alpha and PPAR-gamma controls the transcriptional pathway of adipogenesis and insulin sensitivity. Mol. Cell **3:** 151–158.
62. EL-JACK, A.K., J.K. HAMM, P.F. PILCH, et al. 1999. Reconstitution of insulin-sensitive glucose transport in fibroblasts requires expression of both PPAR-gamma and C/EBP-alpha. J. Biol. Chem. **274:** 7946–7951.
63. ROSEN, E.D., C.H. HSU, X. WANG, et al. 2002. C/EBP-alpha induces adipogenesis through PPAR-gamma: a unified pathway. Genes Dev. **16:** 22–26.
64. TONTONOZ, P., J.B. KIM, R.A. GRAVES, et al. 1993. ADD1: a novel helix-loop-helix transcription factor associated with adipocyte determination and differentiation. Mol. Cell. Biol. **13:** 4753–4739.
65. KIM, J.B. & B.M. SPIEGELMAN. 1996. ADD1/SREBP1 promotes adipocyte differentiation and gene expression linked to fatty acid metabolism. Genes Dev. **10:** 1096–1107.
66. FAJAS, L., K. SCHOONJANS, L. GELMAN, et al. 1999. Regulation of peroxisome proliferator-activated receptor gamma expression by adipocyte differentiation

and determination factor 1/sterol regulatory element binding protein 1: implications for adipocyte differentiation and metabolism. Mol. Cell. Biol. **19**: 5495–5503.
67. KIM, J.B., H.M. WRIGHT, M. WRIGHT, et al. 1998. ADD1/SREBP1 activates PPAR-gamma through the production of endogenous ligand. Proc. Natl. Acad. Sci. USA **95**: 4333–4337.
68. KIM, J.B., P. SARRAF, M. WRIGHT, et al. 1998. Nutritional and insulin regulation of fatty acid synthetase and leptin gene expression through ADD1/SREBP1. J. Clin. Invest. **101**: 1–9.
69. BRUN, R.P., P. TONTONOZ, B.M. FORMAN, et al. 1996. Differential activation of adipogenesis by multiple PPAR isoforms. Genes Dev. **10**: 974–984.
70. HANSEN, J.B., H. ZHANG, T.H. RASMUSSEN, et al. 2001. Peroxisome proliferator-activated receptor delta (PPAR-delta)-mediated regulation of preadipocyte proliferation and gene expression is dependent on cAMP signaling. J. Biol. Chem. **276**: 3175–3182.
71. BASTIE, C., D. HOLST, D. GAILLARD, et al. 1999. Expression of peroxisome proliferator-activated receptor PPAR-delta promotes induction of PPAR-gamma and adipocyte differentiation in 3T3C2 fibroblasts. J. Biol. Chem. **274**: 21920–21925.
72. BASTIE, C., S. LUQUET, D. HOLST, et al. 2000. Alterations of peroxisome proliferator-activated receptor delta activity affect fatty acid-controlled adipose differentiation. J. Biol. Chem. **275**: 38768–38773.
73. BARAK, Y., D. LIAO, W. HE, et al. 2002. Effects of peroxisome proliferator-activated receptor delta on placentation, adiposity, and colorectal cancer. Proc. Natl. Acad. Sci. USA **99**: 303–308.
74. PETERS, J.M., S.S. LEE, W. LI, et al. 2000. Growth, adipose, brain, and skin alterations resulting from targeted disruption of the mouse peroxisome proliferator-activated receptor beta(delta). Mol. Cell. Biol. **20**: 5119–5128.
75. CEDERBERG, A., L.M. GRONNING, B. AHREN, et al. 2001. FOXC2 is a winged helix gene that counteracts obesity, hypertriglyceridemia, and diet-induced insulin resistance. Cell **106**: 563–573.
76. TONG, Q., G. DALGIN, H. XU, et al. 2000. Function of GATA transcription factors in preadipocyte-adipocyte transition. Science **290**: 134–138.
77. SABATAKOS, G., N.A. SIMS, J. CHEN, et al. 2000. Overexpression of Delta-FosB transcription factor(s) increases bone formation and inhibits adipogenesis. Nat. Med. **6**: 985–990.
78. POND, C.M. & C.A. MATTACKS. 1998. In vivo evidence for the involvement of the adipose tissue surrounding lymph nodes in immune responses. Immunol. Lett. **63**: 159–167.
79. POND, C.M. & C.A. MATTACKS. 2002. The activation of the adipose tissue associated with lymph nodes during the early stages of an immune response. Cytokine **17**: 131–139.
80. MATTACKS, C.A. & C.M. POND. 1999. Interactions of noradrenalin and tumour necrosis factor-alpha, interleukin 4 and interleukin 6 in the control of lipolysis from adipocytes around lymph nodes. Cytokine **11**: 334–346.
81. MATTACKS, C.A. & C.M. POND. 1997. The effects of feeding suet-enriched chow on site-specific differences in the composition of triacylglycerol fatty acids in adipose tissue and its interactions in vitro with lymphoid cells. Br. J. Nutr. **77**: 621–643.

82. DESREUMAUX, P., O. ERNST, K. GEBOES, *et al.* 1999. Inflammatory alterations in mesenteric adipose tissue in Crohn's disease. Gastroenterology **117:** 73–81.
83. ROCKSON, S.G. 2001. Lymphedema. Am. J. Med. **110:** 288–295.
84. NOUGUES, J., Y. REYNE & J.P. DULOR. 1988. Differentiation of rabbit adipocyte precursors in primary culture. Int. J. Obes. **12:** 321–333.

A Stepwise Model of the Development of Lymphatic Vasculature

GUILLERMO OLIVER AND NATASHA HARVEY

Department of Genetics, St. Jude Children's Research Hospital, Memphis, Tennessee 38105, USA

ABSTRACT: Although lymphedema was first described more than a century ago, little progress has been made in understanding the mechanisms that cause it. Investigation of the normal development of the lymphatic system has been hindered by the lack of known lymphatic-specific markers. In 1902, F. Sabin proposed the most widely accepted theory about the origin of the lymphatic vasculature. This model proposed that isolated primitive lymph sacs bud from the endothelium of veins during early development; from these primary lymph sacs, the peripheral lymphatic system spreads by endothelial sprouting into tissues where local capillaries form. In 1999, we identified the homeobox gene Prox1 as the first specific marker of lymphatic endothelial cells. Functional inactivation of Prox1 in mice demonstrated that lymphangiogenesis requires the activity of this gene in a subpopulation of endothelial cells in embryonic veins. Prox1 promotes the development of the lymphatic vasculature by determining the final lymphatic fate of budding venous endothelial cells. On the basis of our findings, we propose a stepwise model of lymphangiogenesis in which lymphatic vasculature development is initiated by the specific expression of Prox1 in a subpopulation of vascular endothelial cells that subsequently adopt a lymphatic vasculature phenotype.

KEYWORDS: lymphangiogenesis; Prox1; homeobox; mice; LYVE-1; VEGFR-3

INTRODUCTION

The lymphatic system is a vascular network of thin-walled capillaries and larger vessels lined by a continuous layer of endothelial cells. Lymph vessels drain lymph from the tissue spaces of most organs and return it to the venous system for recirculation. The lack of specific markers has made it difficult to elucidate the mechanism of lymphatic system development. Therefore, the origin of the lymphatic vessels has remained controversial. Historically, the most widely accepted theory of lymphatic development was originally proposed by Sabin.[1,2] On the basis of findings from dye-injection experiments, Sabin proposed that early in development, isolated primitive lymph sacs originate from endothelial cells that bud from the veins. From these primary lymph sacs, the peripheral lymphatic system spreads by endothelial sprouting into the surrounding tissues and organs where local capillaries form.[1–3] In 1910, Huntington and McClure proposed an alternative model, suggesting that the initial lymph sacs arise in mesenchyme, independent of the veins, and subsequently establish venous connections.[4]

The homeobox gene *Prox1* was originally identified on the basis of its homology with the *Drosophila melanogaster* gene prospero.[5] Analysis of the expression pattern of *Prox1* suggested that it has a functional role in various tissues, including lens, retina, liver, pancreas, and central nervous system.[5] Functional inactivation of the *Prox1* gene, which was performed in mice by using an in-frame insertion of the β-galactosidase (β-gal) gene in the *Prox1* locus, leads to embryonic lethality,[6] and phenotypic alterations of the lens and liver in Prox1-null mice have been demonstrated.[6,7]

We have shown that *Prox1* is a specific marker of a subpopulation of endothelial cells that, by budding and sprouting, gives rise to the murine lymphatic system.[8] These findings confirmed Sabin's proposal of the venous origin of the primary lymph sacs. Analysis of *Prox1*-nullizygous embryos revealed that Prox1 activity is essential for the normal development of the lymphatic system; Prox1 activity is required for the budding of the venous endothelial cells and for the determination of their lymphatic fate.[8–10]

LYMPHATIC VASCULATURE DEVELOPS IN A STEPWISE MANNER

To determine the mechanisms by which *Prox1* regulates the budding and sprouting of lymphatic endothelial cells, we initially compared the expression of Prox1 during early murine embryonic development with that of other recently identified lymphatic markers.

The first indication of the initiation of lymphangiogenesis was the specific expression of Prox1 in a restricted subpopulation of endothelial cells in the anterior cardinal vein at embryonic day (E) 9.5.[8] At E10.5, the restricted localization of Prox1 in the veins is still evident, and the first lymphatic endothelial cells have started to bud in a polarized manner (FIG. 1).[8,9] One of the first lymphatic-specific markers to be identified was the vascular endothelial growth factor receptor-3 (*VEGFR-3,* also known as *Flt4*). Early during embryogenesis, VEGFR-3 is expressed in the developing venous endothelium

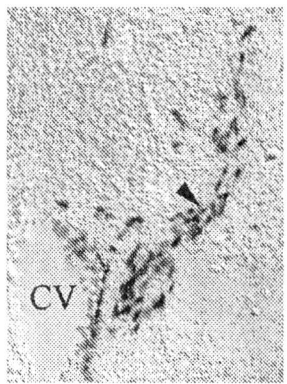

FIGURE 1. Prox1 is a specific marker of a subpopulation of lymphatic precursor endothelial cells in the cardinal vein. At E10.5, Prox1 expression is detected in a subset of venous endothelial cells located near the outer margin of the cardinal vein (CV) in *Prox1*-heterozygous embryos. Initiation of budding of lymphatic endothelial cells proceeds in a polarized manner (*arrowhead*).

and in the presumptive lymphatic endothelial cells; in adult tissues, the expression of VEGFR-3 is mostly restricted to the lymphatic endothelium.[8,11,12] As embryonic development proceeds, the expression of the gene that encodes VEGFR-3 becomes mostly restricted to the lymphatic vessels; however, a low level of expression persists in blood vessels.[8,11] At E10.5, a high level of VEGFR-3 expression was detected in the budding Prox1$^+$ endothelial cells, and a less pronounced level was detected in the arteries and veins.[9]

The lymphatic endothelial hyaluronan receptor (LYVE-1), a CD44 homologue, was recently identified as a specific cell surface protein of lymphatic endothelial cells and macrophages.[13–15] Immunolabeling with antibodies against LYVE-1 and the blood vascular markers PAL-E[16] and CD34[14] revealed that these markers exhibit mutually exclusive expression patterns. In E10.5 budding venous endothelial cells, the expression pattern of Prox1 resembles that of LYVE-1.[9] At this stage, the only difference between the expression patterns of the two markers is that LYVE-1 is uniformly expressed in the endothelial cells of the cardinal vein, and Prox1 expression is polarized and detected only in a restricted subpopulation of endothelial cells.[9]

At E12.5, the number of Prox1- and LYVE-1-positive cells adjacent to the cardinal vein had clearly increased, but Prox1 and LYVE-1 expression was no longer detected in venous endothelial cells.[9] In the lymphatic endothelial cells, the level of VEGFR-3 remained high; however, its expression in vascular endothelial cells was substantially diminished. Maintenance of a high level of VEGFR-3 expression in the lymphatic endothelial cells, its reduced expression level in the vascular endothelial cells, and the initiation of expression of the secondary lymphoid chemokine (SLC; also known as 6Ckine/Exodus-2/CCL21)[17–19] that is released by the lymphatic endothelium in the Prox1$^+$ and LYVE-1$^+$ endothelial cells that have already budded from the veins probably indicate that these budding cells are committed (biased) lymphatic precursors.[9,10]

At E14.5, the lymphatic vasculature had further spread throughout the embryo, and Prox1-positive endothelial cells co-expressed high levels of VEGFR-3; Prox1-negative endothelial cells expressed low levels of this receptor, a finding that indicates that Prox1-negative cells are blood vascular endothelial cells. The pattern of expression of LYVE-1 in lymphatic endothelial cells is similar to that of Prox1; however, LYVE-1 is also expressed in scattered non lymphatic endothelial cells that correspond to macrophages.

PROX1 FUNCTION IS REQUIRED TO SPECIFY THE LYMPHATIC ENDOTHELIAL CELL PHENOTYPE

In E10.5 *Prox1*-nullizygous embryos, a normal number of lymphatic endothelial cell precursors bud from the veins; however, at around E11.5, fewer than normal budding Prox1$^+$ endothelial cells are detected.[8] As expected, the β-gal$^+$ endothelial cells (Prox1-expressing cells) in the *Prox1*-heterozygous embryos also express a high level of VEGFR-3. Only weak VEGFR-3 expression is observed in the budding endothelial cells of the *Prox1*-null littermates, a finding that indicates that those cells are probably blood vasculature endothelial cells.[9] In E11.5 *Prox1*-heterozygous embryos, LYVE-1 is weakly expressed in the cardinal vein and strongly expressed in the budding β-gal$^+$ endothelial cells.[9] The pattern of expression of LYVE-1 overlaps with that of β-gal in the cardinal vein, but not with that in the budding β-gal$^+$ endothelial cells of *Prox1*-nullizygous littermates.[9]

In contrast to the lymphatic vessels, vascular vessels exhibit a distinct, continuous basal membrane that contains laminin.[20] In addition, blood endothelial cells express the surface glycoprotein CD34[21]; lymphatic endothelial cells do not, or only at very low levels. In E11.5 *Prox1*-heterozygous embryos, β-gal$^+$ endothelial cells budding from the cardinal vein do not coexpress laminin or CD34; in the *Prox1*-null littermates, budding cells express high levels of laminin and CD34.[9] These results indicate that the budding endothelial cells in E11.5-12.0 *Prox1*-null embryos have adopted a vascular phenotype rather than a lymphatic one.

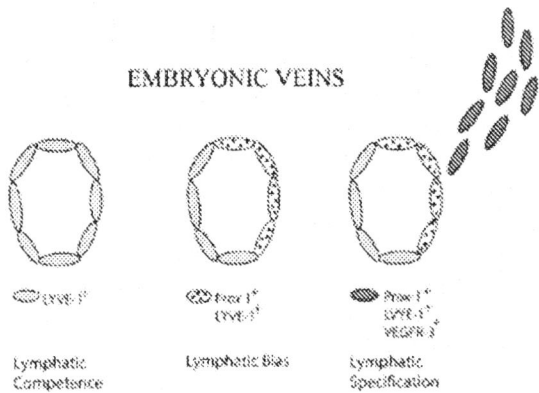

FIGURE 2. Scheme of the model proposed for the embryonic development of the mammalian lymphatic vasculature. Following the formation of the blood vascular system, LYVE-1 starts to be expressed in venous endothelial cells in the cardinal vein at approximately E9.5 to E10.0. At this stage a subpopulation of venous endothelial cells becomes competent to respond to a specific lymphatic inductive signal (lymphatic competence), which results in the induction of Prox1 expression in a restricted subpopulation of endothelial cells in the cardinal vein. Most likely, Prox1 expression in these cells initiates the lymphatic differentiation program in which those Prox1-expressing venous endothelial cells become biased (committed) to a lymphatic phenotype. Expression of LYVE-1 and Prox1 is one of the first indications that lymphangiogenesis has been initiated (lymphatic bias). As development proceeds, this subpopulation of LYVE-1- and Prox1-positive endothelial cells starts to bud from the veins in a Prox1-independent manner. At this stage VEGFR-3 expression still remains almost equally high in both vascular and lymphatic endothelial cells. As the cells bud from the veins in a polarized manner, they start to express additional lymphatic endothelial markers. At around E11.5, VEGFR-3 expression level remains high in budding lymphatic endothelial cells, but it weakens in blood vascular endothelial cells. At this stage lymphangiogenesis is irreversible (lymphatic specification).

These results suggest that Prox1 activity is required not only to maintain budding and sprouting of a subpopulation of venous endothelial cells that will give rise to the lymphatic vasculature, but also to determine the final fate of those budding endothelial cells. On the basis of these results, we developed a working model of early lymphatic vasculature development (FIG. 2). In this model and following the initial formation of the blood vascular system, LYVE-1 and Prox1 are two of the earliest markers expressed in endothelial cells in the cardinal vein at approximately E9.5 to E10.0; the detection of these markers is one of the first indications that lymphangiogenesis has been initiated. We propose that the expression of Prox1 in a restricted subpopulation of venous endothelial cells (precursors) is sufficient to promote lymphatic development, and its expression initiates the commitment to the lymphatic differentiation program. As development proceeds, this subpopulation of LYVE-1- and Prox1-positive endothelial cells buds from the veins in an initially Prox1-independent manner. However, maintenance of budding requires Prox1 activity. Normally, as the cells bud in a polarized manner, they start to express additional lymphatic endothelial markers in a stepwise fashion. The expression of those lymphatic markers may indicate that this process becomes irreversibly biased toward the lymphatic pathway. Therefore, Prox1 activity in a restricted subpopulation of endothelial cells in the embryonic veins is probably required not only to promote lymphangiogenesis, but also to determine the lymphatic fate of the differentiation program of those budding venous endothelial cells.

ACKNOWLEDGMENTS

This work is supported by Grants GM58462 and EY12162 from the National Institutes of Health, by Cancer Center Support (CORE) Grant CA21765 from the National Cancer Institute, and by the American Lebanese Syrian Associated Charities (ALSAC).

REFERENCES

1. SABIN, F.R. 1902. On the origin of the lymphatic system from the veins, and the development of the lymph hearts and thoracic duct in the pig. Am. J. Anat. **1**: 367–389.
2. SABIN, F.R. 1904. On the development of the superficial lymphatics in the skin of the pig. Am. J. Anat. **3**: 183–195.
3. GRAY, H. 1985. The lymphatic system. *In* Anatomy of the Human Body. C.D. Clemente, Ed. :866–932. Lea and Febiger. Philadelphia, PA.
4. HUNTINGTON, G.S. & C.F.W. McClure. 1910. The anatomy and development of the jugular lymph sac in the domestic cat (*Felis domestica*). Am. J. Anat. **10**: 177–311.
5. OLIVER, G., B. SOSA-PINEDA, S. GEISENDORF, *et al.* 1993. Prox 1, a prospero-related homeobox gene expressed during mouse development. Mech. Dev. **44**: 3–16.
6. WIGLE, J.T., K. CHOWDHURY, P. GRUSS & G. OLIVER. 1999. *Prox1* function is crucial for mouse lens-fibre elongation. Nat. Genet. **21**: 318–322.
7. SOSA-PINEDA, B., J.T. WIGLE & G. OLIVER. 2000. Hepatocyte migration during liver development requires *Prox1*. Nat. Genet. **25** 254–255.
8. WIGLE, J.T. & G. OLIVER. 1999. *Prox1* function is required for the development of the murine lymphatic system. Cell **98**: 769–778.

9. WIGLE, J.T., N. HARVEY, M. DETMAR, et al. 2002. An essential role for Prox1 in the induction of the lymphatic endothelial cell phenotype. EMBO J. **21:** 1505–1513.
10. OLIVER, G. & M. DETMAR. 2002. The rediscovery of the lymphatic system: new and old insights into the development and biological function of the lymphatic vasculature. Genes & Devel. **16:** 773–783.
11. KAIPAINEN, A., J. KORHONEN, T. MUSTONEN, et al. 1995. Expression of the fms-like tyrosine kinase 4 gene becomes restricted to lymphatic endothelium during development. Proc. Natl. Acad. Sci. USA. **92:** 3566–3570.
12. VALTOLA, R., P. SALVEN, P. HEIKKILA, et al. 1999. VEGFR-3 and its ligand VEGF-C are associated with angiogenesis in breast cancer. Am. J. Pathol. **154:** 1381–1390.
13. BANERJI, S., J. NI, S.X. WANG, et al. 1999. LYVE-1, a new homologue of the CD44 glycoprotein, is a lymph-specific receptor for hyaluronan. J. Cell Biol. **144:** 789–801.
14. PREVO, R., S. BANERJI, D.J. FERGUSON, et al. 2001. Mouse LYVE-1 is an endocytic receptor for hyaluronan in lymphatic endothelium. J. Biol. Chem. **276:** 19420–19430.
15. JACKSON, D.G., R. PREVO, S. CLASPER & S. BANERJI. 2001. LYVE-1, the lymphatic system and tumor lymphangiogenesis. Trends Immunol. **22:** 317–321.
16. SKOBE, M. & M. DETMAR. 2000. Structure, function and molecular control of the skin lymphatic system. J. Invest. Dermatol. Symp. Proc. **5:** 14–19.
17. ZLOTNIK, A. & O. YOSHIE. 2000. Chemokines: a new classification system and their role in immunity. Immunity **12:** 121–127.
18. GUNN, M.D., K. TANGEMANN, C. TAM, et al. 1998. A chemokine expressed in lymphoid high endothelial venules promotes the adhesion and chemotaxis of naive T lymphocytes. Proc. Natl. Acad. Sci. USA **95:** 258–263.
19. GUNN, M.D., S. KYUWA, C. TAM, et al. 1999. Mice lacking expression of secondary lymphoid organ chemokine have defects in lymphocyte homing and dendritic cell localization. J. Exp. Med. **189:** 451–460.
20. EZAKI, T., K. MATSUNO, H. FUJII, et al. 1990. A new approach for identification of rat lymphatic capillaries using a monoclonal antibody. Arch. Histo.l Cytol. **53 (Suppl):** 77–86.
21. PAAL, E., L.D. THOMPSON & C.F. HEFFESS. 1998. A clinicopathologic and immunohistochemical study of ten pancreatic lymphangiomas and a review of the literature. Cancer **82:** 2150–2158.

De Novo Lymph Node Formation in Chronic Inflammation of the Human Leg

WALDEMAR L. OLSZEWSKI

Department of Surgical Research and Transplantology, Medical Research Center, Polish Academy of Sciences, 02 106 Warsaw, Poland

ABSTRACT: Organized lymphoid tissue is the first line of antigenic defense. Recruited by antigen located in the non-lymphoid tissues, the infiltrating lymphocytes often organize themselves as follicle-like structures that contain germinal centers, similar to those in secondary lymphoid follicles of lymph nodes. These extranodal tertiary lymphoid follicles are found in various autoimmune diseases. We investigated 153 patients with protracted lymph stasis of the lower limb, caused by lymphatic damage incurred through soft tissue bacterial inflammation or mechanical trauma of soft tissues and bones. In 10% of patients with post-inflammatory, and in 25% with post-traumatic lymph stasis, "newly-formed" lymph nodes were detected by means of lymphoscintigraphy. They were located along the large calf and thigh veins. Although scattered nodes are normally detected in these areas, the number and total mass of visualized nodes substantially exceeded those seen in healthy subjects. The calculated surface area of "newly formed" nodes attained 50–70% of the area of ipsilateral inguinal nodes. Histological evaluation of nodal biopsy specimens in three such patients revealed, in one, a lymph node structure without differentiation into cortical and medullary areas, and in a second, a follicle-like structure within a dilated lymph vessel. Lymph clot removed from another dilated vessel contained a lymphocyte/dendritic cell aggregate. The "newly formed" nodes likely originate from primordial lymphoid follicles and/or lymphoid cell aggregates formed in response to chronic stimulation by microbial products and self-antigens from the damaged tissues. Detection of "newly-formed" lymph nodes in the limb is evidence of an ongoing inflammatory process and requires appropriate therapy.

KEYWORDS: lymph nodes; neogenesis; infection; trauma

Address for correspondence: W.L. Olszewski, M.D., Ph.D., Medical Research Center, Pawinski Str. 5, 02106 Warsaw, Poland. Voice: +48226685316; fax: +48226685334.
wlo@cmdik.pan.pl

INTRODUCTION

Lymphoid infiltrates often organize themselves as follicle-like structures that contain germinal centers similar to those in secondary lymphoid follicles of lymph nodes. Ectopic or extranodal tertiary lymphoid follicles are found in autoimmune diseases, such as rheumatoid synovium,[1,2] the thymus of myasthenia gravis,[3] the salivary glands of Sjogren's disease,[4] the liver in hepatitis C,[5,6] autoimmune thyroid diseases,[7] chronic inflammatory bowel disease,[8] and *Helicobacter pylori*–infected gastric mucosa.[9] The cell proliferation and apoptosis of the ectopic lymphoid structures resemble those of sustained germinal center reactions. Germinal centers can be formed *de novo* in non-lymphoid tissues through the process of lymphoid neogenesis.[10] Homing chemokines and high endothelial venules participate in cell recruitment and compartmentalization into functional zones. The question arises as to which factors ultimately control the process. A critical element is antigen. It is suggested that the antigens to which the new germinal centers are committed are components of the host's own tissues. However, there is evidence that the B cells that are incorporated into the lymphoid structures associated with infection are specific for bacterial antigens. Thus, local chronic infections in the non-lymphoid tissues may also be responsible for formation of organized lymphoid infiltrates.

We have observed the development of newly organized lymphoid structures in the lower limbs of persons with post-inflammatory and post-traumatic lymph stasis by means of isotopic lymphography. These structures were located along the lymphatics that drain chronically infected soft tissues or mechanically injured soft tissues and bones.

Until recently, lymphoscintigraphy has been exclusively utilized to delineate lymphatic pathways and to evaluate the dynamics of lymph flow in cases of clinically diagnosed lymphedema. We have introduced this method as a routine diagnostic procedure for all cases of lower limb edema of non-systemic etiology. This method has proven to be useful not only for the delineation of lymphatic pathways, but also in the detection of inflammatory infiltrates, enlarged deep lymph nodes, and newly formed lymphoid cell aggregates.

MATERIAL AND METHODS

Patients

One hundred and fifty-three randomly selected patients with long-standing post-inflammatory and post-traumatic edema of the leg were included in the study. All of these patients were referred for evaluation of foot and calf edema complicated by recurrent incidents of dermatolymphangioadenitis or chronic inflammation of the soft tissues of the calf. In 110 cases, the edema was preceded historically by an episode of acute cutaneous inflammation in the limb (Group 1). In 43 cases, the development of edema was linked to soft tissue or bony injury (injury of muscle or Achilles tendon, ankle joint dislocation, or fracture of the tibia, femur, or calcaneus) and was followed by protracted local inflammation at the site of trauma (Group 2). In Group 1, no changes were detected in the large veins, whereas in Group 2, post-thrombotic venous alterations in the calf or thigh were observed in 21% of cases. None of the patients had ulceration, open wounds, fistulae, osteomyelitis, or primary lymphedema, either in the past or at the time of inclusion into the study. An informed consent for participation in the study was signed by each patient and was approved by the Institute's Ethical Committee.

Lymphoscintigraphic Investigation

Lymphoscintigraphy was performed in both legs 30, 90, and 150 minutes after subcutaneous injection of ^{99m}Tc-Nanocol (3 mCi) into the first web space. A gamma camera (Orbiter ZLC 750, Siemens, Germany) was utilized. The speed of visualization of calf and thigh lymphatics and filling of inguinal and iliac lymph nodes was qualitatively evaluated (< or > 30 minutes). All abnormalities (for example, dermal backflow of the tracer, collateral circulation, local accumulation of tracer in non-lymphoid tissues and ectopic lymph nodes) were recorded. For quantitative purposes, the lymphoscintigrams were scanned and analyzed using specialized PC software (Olympus Micro Image™ version 3.0.0., Olympus Optical Co., Hamburg, Germany). The surface area of the calf and thigh "newly formed" nodes and inguinal lymph nodes of both legs was measured. Data were expressed as indices, obtained from the equations $I_{NLN} = S_{NLN}/S_{INGLNe}$ and $I_{INGLN} = S_{INGLNe}/S_{INGLNc}$, where S_{NLN} was the surface of new lymph nodes and S_{INGLN} the surface of inguinal nodes measured in the edematous (e) and contralateral (c) extremity.

Lymph Node Biopsy

In three cases in Group 1, biopsy of the "newly formed" lymph nodes was performed at the time of microsurgical lympho-venous shunt operation. The specimens were fixed in formalin and sections were stained with hematoxylin–eosin.

RESULTS

Group 1 (Postinflammatory Lymph Stasis)

Lymphoscintigraphic Evaluation

In 11 of 110 patients (10%) "newly formed" lymph nodes were visualized in the popliteal region (9 cases), thigh (1 case), and calf (1 case) (FIG. 1). In 4

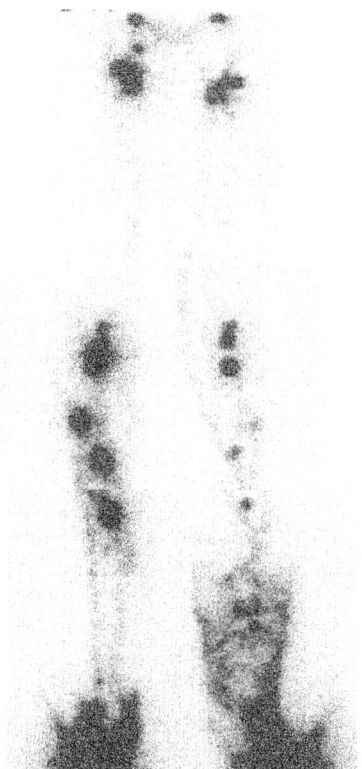

FIGURE 1. Lymphoscintigram of a patient with post-inflammatory lymph stasis in the lower limbs. Multiple "newly formed" lymph nodes are seen along the path of the deep lymphatics in both limbs.

of 11 patients with bilateral edema, "newly formed" lymph nodes were seen in both limbs. There were 1 to 5 nodes in each swollen limb (mean 3.45/limb). The surface area index I_{NLN} ("newly formed"/ipsilateral lymph nodes) was 0.56 ± 0.3, and I_{INGLN} (edematous limb inguinal/contralateral inguinal) was 1.05 ± 1.0 (FIG. 2). All lymphoscintigrams revealed signs of lymph stasis.

Clinical Evaluation of Patients with "Newly Formed" Lymph Nodes

The mean duration of edema was 8.75 ± 7 (range, 0.5–20) years. All patients had 1–5 recurrent attacks of dermatolymphangioadenitis per year. In the latent periods between attacks, the skin over the swollen limbs remained warmer than the non-affected contralateral extremity, and was slightly erythematous. Broad-spectrum antibiotics were administered during acute attacks and short-term manual lymph drainage was employed.

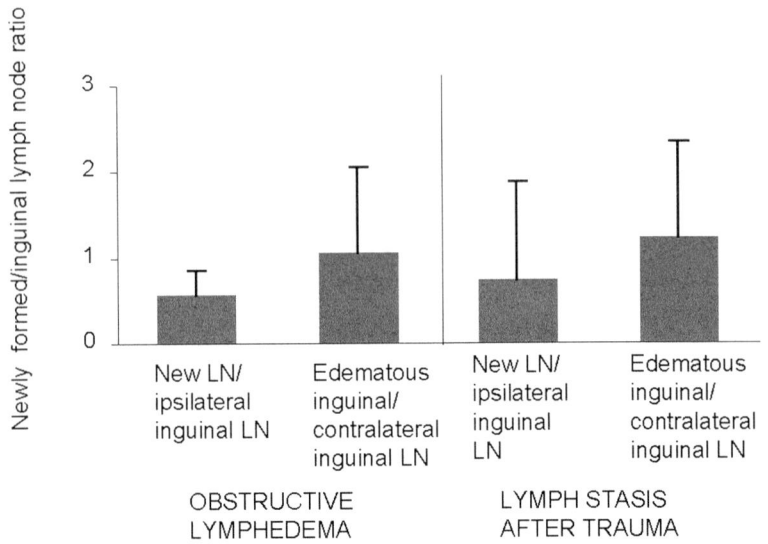

FIGURE 2. Surface area of the lymphoscintigraphic image of "newly formed" lymph nodes. The data are presented as indices (ratios) of the surface areas of "new" to ipsilateral inguinal nodes. The surface areas of inguinal lymph nodes from both sides were compared. The surface (mass) of "new" nodes attained 50–70% of inguinal node values. There were no differences in the mass of inguinal nodes between the swollen and the healthy limb.

Group 2 (Post-traumatic Lymph Stasis)

Lymphoscintigraphic Evaluation

In 11 of 43 patients (25%) "newly formed" lymph nodes were visualized in the popliteal region (10 cases) and thigh (1 case) (FIG. 3). In 1 of 11 patients with bilateral edema, "newly formed" nodes were seen in both limbs. There were 1 to 5 nodes in each swollen limb (mean 2.72/limb). The I_{NLN} ("newly formed"/ispilateral lymph nodes) was 0.76 ± 1.15 and I_{INGLN} (edematous limb inguinal/contralateral inguinal) was 1.24 ± 1.09 (FIG. 2). All lymphoscintigrams showed dilated lymphatics in the calf and thigh.

Clinical Evaluation of Patients with "Newly Formed" Lymph Nodes

The mean duration of edema was 6 ± 3.9 (range, 2–18) months. The swollen portions of the limb were painful during movement and slightly erythematous, with an elevation of the surface temperature. All patients received non-steroidal anti-inflammatory drugs.

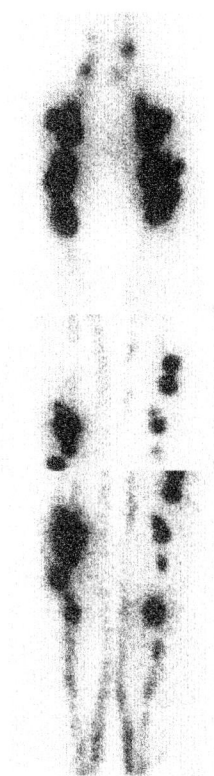

FIGURE 3. Lymphoscintigram of a patient 6 months after injury of the soft tissues in the calves. There was a chronic inflammatory process at the site of injury. Multiple lymph nodes are notable along the deep lymphatics.

Macroscopic Appearance and Histology of "Newly Formed" Lymph Nodes

At the time of lympho-venous shunt surgery, the inguinal lymph nodes and afferent lymphatics were found to be fibrotic and adherent to the fibrotic tissue along the great saphenous vein. Distal to the inguinal nodes, small, single, bluish nodes were detected, and these were observed to bleed upon incision. Occasionally, scattered dilated segments of lymphatics could be seen. The lymph node–like structures and dilated fragments of lymphatics were harvested for histological evaluation, which revealed either lymphoid cell aggregates, separated by fibrous strands, with a thick capsule around the entire structure (FIG. 4) or accumulation of mononuclear cells in the lumina of dilated lymphatics, spatially oriented to resemble organized lymphoid tissue (FIG. 5). In one case, a clot was removed from the dilated lymphatic, containing organized lymphoid structures with lymphocytes and large dendritic cells (FIG. 6).

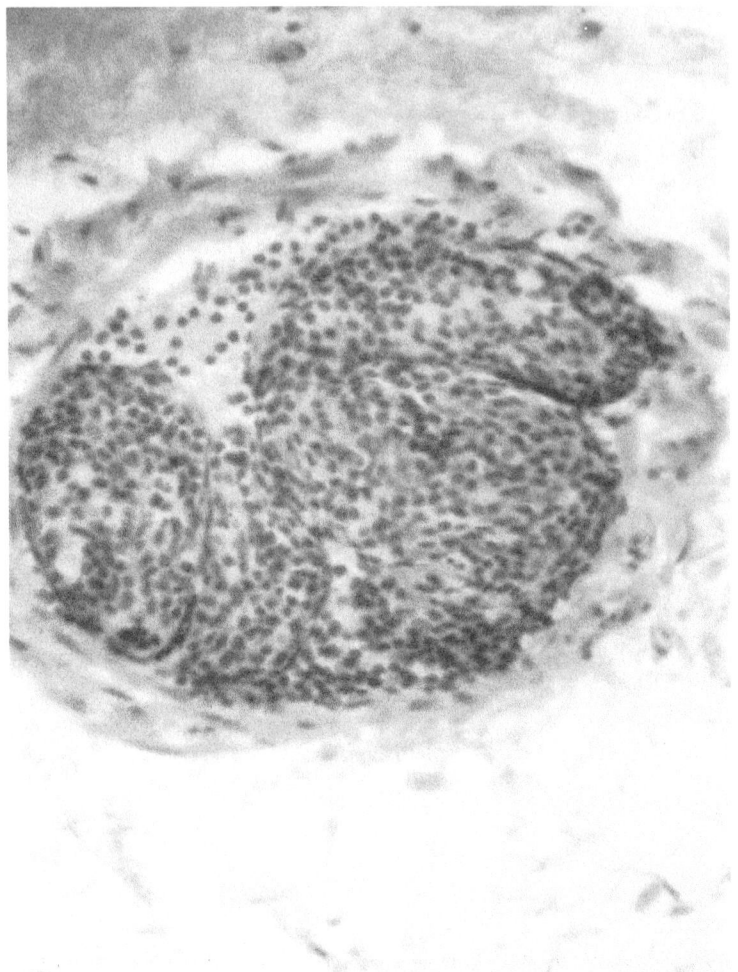

FIGURE 4. The histology of a "newly formed" lymph node from a patient with long-standing lymph stasis of the lower limb. A lymph node with compartments, separated by septa, without differentiation into cortical and medullary areas. A thick capsule (lymph vessel wall?) is seen. (Hematoxylin and eosin stain; original magnification ×400.)

DISCUSSION

In this study, we have performed a lymphoscintigraphic examination of "newly formed" lymph nodes in the lower extremities of patients with long-standing post-inflammatory and post-traumatic lymph stasis. This isotopic method permits the visualization of multiple lymph node structures located along the lymphatics that approximate the popliteal and femoral veins. In

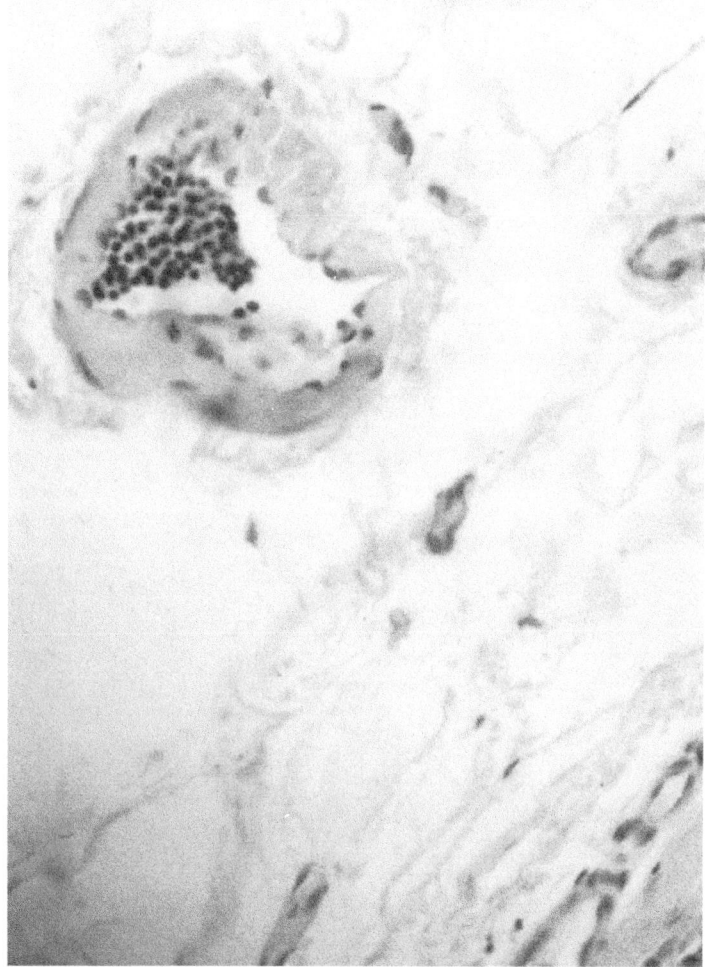

FIGURE 5. An intralymphatic aggregate of lymphoid cells containing lymphocytes and dendritic cells. The structure surrounding the aggregate is the wall of a dilated lymph vessel. (Hematoxylin and eosin stain; original magnification ×200.)

health, these sites typically, on occasion, display only one or two small lymph nodes. The histological evaluation of the "newly formed" lymphoid structures revealed the presence of lymphoid cell aggregates, either in the lumen of dilated lymphatics or surrounded by a capsule with a dilated sinus beneath.

The patients inducted into this study were randomly selected from those presenting with chronic edema and inflammation of the distal lower limb. Analysis of the historical presentation yielded two groups by etiology. In the first, the onset of edema was preceded by inflammation of soft tissues, and,

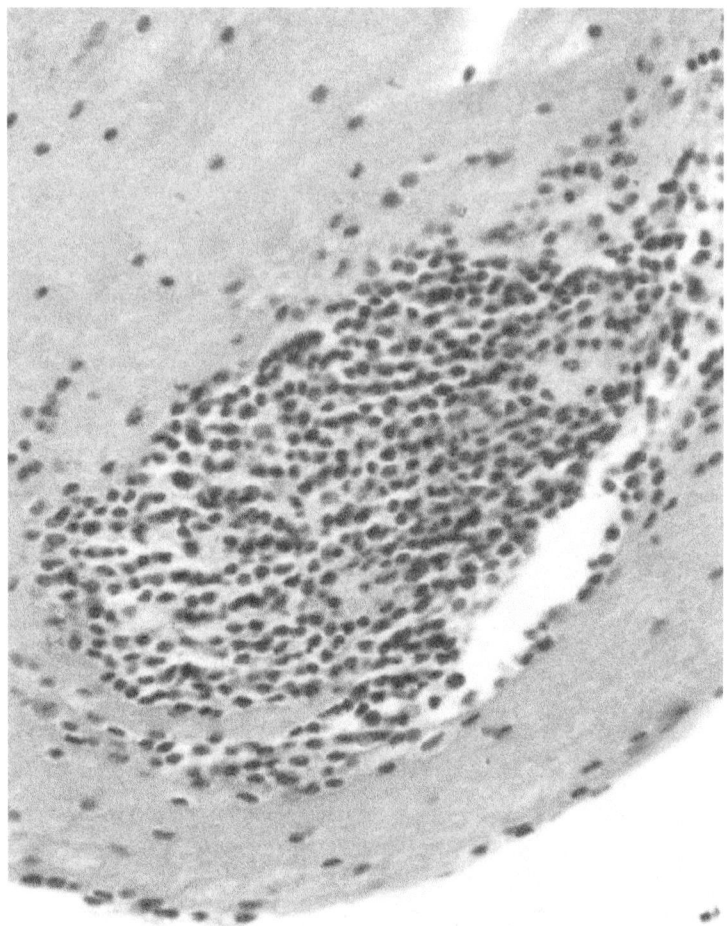

FIGURE 6. A lymphoid cell aggregate in a lymph clot removed form a dilated lymph vessel. Lymphocytes and dendritic cells forming a structure resembling organized lymphoid tissue. (Hematoxylin and eosin stain; original magnification ×400.)

in the second, edema and long-standing inflammation developed after mechanical trauma of soft tissues or bones. The common denominator for both groups was a chronic inflammatory process in foot or calf tissues. Without regard to the etiology, the lymphoscintigraphic response was similar. From the immunological standpoint, the lymphoid reaction of the limb in postinflammatory lymph stasis, expressed as development of new lymph node genesis or enlargement of pre-existing nodes, was likely initiated and maintained by microbial antigens,[11] whereas in the posttraumatic stasis it was presumably caused by self-antigens from the damaged tissues.[12]

The question remains whether the visualized nodes correspond to pre-existing nodes in a primordial stage or that they formed from the recruited lymphocyte aggregates. Previous anatomical[13] and lymphangiographic[14] studies have shown that, under physiological conditions, there may be one or two small popliteal and isolated tibial, fibular, and femoral nodes along large vessels. However, their diameters do not normally exceed 5–7 mm and they are only occasionally delineated on lymphography. Obstruction of superficial lymphatics would redirect the lymph stream to the deep lymphatics that parallel the large arteries and veins. In this circumstance, some deep lymph nodes can be visualized. However, in both patient groups in this study, the superficial lymphatics remained at least partally patent and inguinal nodes were clearly visualized. Moreover, the lymphoscintigraphic total area of the "newly formed" nodes attained 50 to 70% of the surface area of the inguinal nodes. Therefore, it seems quite unlikely that such a mass of lymph nodes would exist in the limb under normal conditions.

Although only three "newly formed" nodes were biopsied, in each case, the histologic picture clearly shows either a node with a primitive structure that lacks follicles, or intralymphatic lymphocyte aggregates. This appearance would seem to favor the process of node formation from an intralympatic aggregate,[15] rather than enlargement of pre-existing nodes, populated by recruited lymphocytes. However, the latter cannot be excluded. We did not carry out immunohistochemical studies to document the presence of chemokines, such as CCL21 and CXCL13, that are responsible for the formation of lymphoid structures and might therefore be responsible for the stimulation of lymph node genesis

There is a growing body of evidence on lymphoid neogenesis in chronic inflammation. The TNF proteins and homing chemokines play an important role in this process. Continuous antigen presentation is probably the main driving force for the formation of organized lymphoid tissue. The question arises as to whether this process is beneficial for the organism, through the elimination of antigens, or whether lymphoid neogenesis is a reflection of an excessive autoimmune reaction. Local bacterial or viral infection or destruction of the host cells may lead to expression of previously unrecognized self-antigens and subsequently bring about an autoimmune reaction.

A series of chemokines and cytokines has been implicated in the process of lymphoid organogenesis, suggesting that the same biological principles are shared in secondary and tertiary lymphoid structures. Among the chemokines, CXCL13 (previously BCL or BCA-1)[16–18] and CCL21 (formerly SLC)[6] have received attention. They are expressed on the high endothelial venule cells.[19] The CCL21 preferentially attracts T cells, whereas CXCL13 is a strong B-cell attractant. An important role in lymphoid organogenesis has been assigned to lymphotoxin LTα or LTβ of the tumor necrosis factor family (TNF) and receptors TNFR-I p55, TNFR-II p 75, and LTβR.[16,20,21]

Ruddle et al.[10] have shown a connection between inflammation and the formation of ectopic lymphoid structures. Introducing an LTα gene resulted in formation of germinal centers, expression of adhesion molecules, and an array of chemokines, including CXCL13 and CCL21.[1] However, there was an absolute necessity for follicular dendritic cells (FDC) to be present, without which the germinal center could not be created. These cells are not represented at extranodal sites. Thus, FDC had to be recruited from bone marrow.[22]

The process of formation of tertiary lymphoid tissue can be schematically described. Antigen initiates cell recruitment and release of TNFα, IL-1, and INFγ by these cells. There are two pathways along which further processes proceed. Chemokines CXCL13, CCL21, and LTβ accumulate; follicular dendritic cell precursors are attracted; and, subsequently, CD8CD40L T cells are recruited. A germinal center is formed. In another pathway T and B cells accumulate, attracted by CCL21 chemokine, and a cellular aggregate is formed.[22]

Our future immunohistochemical studies of lymph node biopsy specimens should permit further investigation of whether the "newly formed" lymph nodes arise from lymphoid cell aggregates or primordial lymphatic follicles.

REFERENCES

1. TAKEMURA, S., A. BRAUN & C. CROWSON. et al. 2001. Lymphoid neogenesis in rheumatoid synovitis. J. Immunol. **167:** 1072–1080
2. RANDEN, I., O.J. MELLBYE, Ø. FØRRE & J.B. NATVIG. 1995. The identification of germinal centres and follicular dendritic cell networks in rheumatoid synovial tissue. Scand. J. Immunol. **41:** 481–486.
3. MURAI, H., H. HARA, T. HATAE, et al. 1997. Expression of CD23 in the germinal center of thymus from myasthenia gravis patients. J. Neuroimmunol. **76:** 61–69.
4. AZIZ, K.E., P.J. MCCLUSKEY & D. WAKEFIELD. 1997. Characterisation of follicular dendritic cells in labial salivary glands of patients with primary Sjogren syndrome: comparison with tonsillar lymphoid follicles. Ann. Rheum. Dis. **56:** 140–143.
5. FRENI, M.A., D. ARTUSO, G. GERKEN, et al. 1995. Focal lymphocytic aggregates in chronic hepatitis C: occurrence, immunohistochemical characterization, and relation to markers of autoimmunity. Hepatology **22:** 389–394.
6. GRANT, A.J., S. GODDARD, J. AHMED-CHOUDHURY, et al. 2002. Hepatic expression of secondary lymphoid chemokine (CCL21) promotes the development of portal-associated lymphoid tissue in chronic inflammatory liver disease. Am. J. Pathol. **160:** 1445–1455.
7. ARMENGOL, M.P., M. JUAN, A. LUCAS-MARTIN, et al. 2001. Demonstration of thyroid antigen-specific B cells and recombination-activating gene expression in chemokine-containing active intrathyroidal germinal centers. Am. J. Pathol. **159:** 816–873.

8. KAISERLING, E. 2001. Newly-formed lymph nodes in the submucosa in chronic inflammatory bowel disease. Lymphology **34:** 22–29.
9. MAZZUCCHELLI, L., A. BLASER, A. KAPPELER, et al. 1999. BCA-1 is highly expressed in *Helicobacter pylori*-induced mucosa-associated lymphoid tissue and gastric lymphoma. J. Clin. Invest. **104:** R49–R54.
10. RUDDLE, N.H. 1999. Lymphoid neo-organogenesis: lymphotoxin's role in inflammation and development. Immunol. Res. **19:** 119–125
11. OLSZEWSKI, W.L., S. JAMAL, G. MANOKARAN, et al. 1997. Bacteriologic studies of ski, tissue fluid, lymph and lymph nodes in patients with filarial lymphedema. Am. J. Trop. Med. Hyg. **57:** 7–15.
12. SZCZESNY, G. & W.L. OLSZEWSKI. 2002. The pathomechanism of posttraumatic edema of lower limbs. I. The effect of extravasated blood, bone marrow cells, and bacterial colonization on tissues, lymphatics and lymph nodes. J. Trauma **52(2):** 315–322.
13. KUBIK, S. 2000. Atlas of the Lymphatics of the Lower Limbs. Servier. Paris.
14. KAINDL, F., E. MANNHEIMER, L. PFLEGER-SCHWARY, et al. 1960. Lymphangiographie und Lymphadenographie der unteren Extremitaeten.:2–3. Thieme. Stuttgart.
15. EIKELENBOOM, P., J.J.J. NASSY, J. POST, et al. 1978. The histogenesis of lymph nodes in rat and rabbit. Anat. Rec. **190:** 201–216.
16. HJELMSTRÖM, P., J. FJELL, T. NAKAGAWA, et al. 2000. Lymphoid tissue homing chemokines are expressed in chronic inflammation. Am. J. Pathol. **156:** 1133–1138.
17. LEGLER, D.F., M. LOETSCHER, R. STUBER ROOS, et al. 1998. B cell-attracting chemokine 1, a human CXC chemokine expressed in lymphoid tissue, selectively attracts B lymphocytes via BLR1/CXCR5. J. Exp. Med. **187:** 655–660.
18. ANSEL, K.M., V.N. NGO, P.L. HYMAN, et al. 2000. A chemokine-driven positive feedback loop organizes lymphoid follicles. Nature **406:** 309–314.
19. ZLOTNIK, A., J. MORALES & J.A. HEDRICK. 1999. Recent advances in chemokine and chemokine receptors. Crit. Rev. Immunol. **19:** 1–47
20. YANG-XIN, F. & D.D. CHAPLIN. 1999. Development and maturation of secondary lymphoid tissues. Annu. Rev. Immunol. **17:** 399–433
21. FÜTTERER, A., K. MINK, A. LUZ, et al. 1998. The lymphotoxin β receptor controls organogenesis and affinity maturation in peripheral lymphoid tissues. Immunity **9:** 59–70.
22. WEYAND C., P.J. KURTIN & J.J. GORONZY. 2001. Ectopic lymphoid organogenesis. Am. J. Pathol. **159:** 787–793.

Physiologic Aspects of Lymphatic Contractile Function

Current Perspectives

ANATOLIY A. GASHEV

Department of Medical Physiology, Cardiovascular Research Institute, College of Medicine, Texas A&M University System Health Science Center, College Station, Texas 77843-1114, USA

ABSTRACT: The lymphatic system plays an important role in fluid/macromolecular balance, lipid absorption, and immune functions, and is involved in many different pathologic conditions, like inflammation, spread of cancer cells, and lymphedema. There are several forces that drive lymph centripetally. Extrinsic driving forces, or the passive lymph pump, include lymph formation, arterial pulsations, skeletal muscles contractions, fluctuations of central venous pressure, gastrointestinal peristalsis, and respiration. Intrinsic forces, or the active lymph pump, are the result of coordinated contractions of lymphangions, the morpho-functional units of the lymphatic vessels, which include the valve and portion of the vessel extending to the next valve. The contractions of the lymphangions are initiated by the pacemaker activity of the smooth muscle cells of lymphangion wall. Transmural pressure is an important hydrodynamic factor that modulates pacemaking. Under conditions of low filling, lymphangions might produce negative intraluminal pressures and a suction effect. Because of the complicated hydrodynamic conditions in lymphatic beds, the passive and active lymph pumps sometimes work together to propel lymph centripetally. In other cases (i.e., under conditions of enhanced lymph flow), flow-mediated inhibition of the active lymph pump could serve to decrease lymphatic outflow resistance and save metabolic energy when the driving force of the passive lymph pump is enough to propel lymph. We have recently found that there are profound differences in the pressure and flow sensitivities of lymphatic vessels derived from different tissues, such as the thoracic duct and mesenteric lymphatics. Such results, when considered in light of the controversy surrounding some studies performed in different animals, lead to the idea that the active lymph pumps in humans may have greater regional differences in contractile function than has been seen in

---

Address for correspondence: Dr. Anatoliy Gashev, Research Assistant Professor, Department of Medical Physiology, Cardiovascular Research Institute, College of Medicine, Texas A&M University System Health Science Center, 336 Reynolds Medical Building, College Station, TX 77843-1114, USA. Voice: 979-862-8575; fax: 979-847-8635.

gashev@tamu.edu

animals, because of the upright posture in bipedal humans. This posture creates an additional outflow resistance for lymphatics of the lower part of the body. Thus, despite the ongoing attempts to determine the mechanisms of lymphatic diseases and useful therapies to treat them, there are many disputable or unknown issues regarding the physiology of lymph transport in humans.

KEYWORDS: lymphatics; lymphangion; lymph pressure; lymph flow; lymph pump

The lymphatic transport system plays significant roles in the maintenance of body fluid and macromolecular balances, in lipid absorption, and in immune reactions. Damage to the transport capabilities of the lymphatic system causes lymphedema, which is connected with different pathologic conditions, such as inflammation, invasion of parasites or bacteria, and partial or full occlusion of lymphatic vessels and nodes after surgical manipulations or irradiation of tumors. In spite of ongoing attempts to discover the mechanisms of and treatments for lymphatic diseases, there are a many disputable or unknown issues regarding the physiology of lymph transport.

INTRALUMINAL PRESSURE AS A MODULATOR OF ACTIVE LYMPH PUMPS

There are a several forces that drive lymph centripetally. One group is the passive lymph pump, or extrinsic driving forces, such as lymph formation, arterial pulsations, contractions of skeletal muscles, central venous pressure fluctuations, gastrointestinal peristalsis, and respiration. These forces produce passive hydrostatic gradients in the lymphatic network, which could effectively propel lymph, even without the influences of the intrinsic lymphatic contractions. The intrinsic or active lymph pumps are the coordinated contractions of the chains of lymphangions. Lymphangions are the morphological-functional units of the lymphatic vessels, first described by Mislin,[1,2] that include the valve and adjacent portion of the lymphatic vessel extending to the next valve. There are layers of smooth muscle cells in a lymphatic vessel wall and a layer of endothelial cells, which covers the inner surface.

Contractions of the lymphangions are initiated by the pacemaker activity of the smooth muscle cells of lymphangion wall.[3–7] Spontaneous depolarizations cause the generation of the action potentials that result in the contractions of the lymphangion. The coordinated contractions of muscle cells in the lymphangion wall cause local increases of intravascular pressure inside the lymphatic vessels. Local pressure gradients that develop during the contraction lead to the closure of the lymphangion input valve, opening of the output valve. and the centripetal propulsion of lymph.[8–10] The contractile cycle in lymphangions has been described using cardiac analogies and divided into periods of diastole and systole.[8]

Transmural pressure is an important hydrodynamic factor. It influences the contractile activity of lymphangions, causing inotropic effects (changes in the strength of contraction) and chronotropic effects (changes in the contraction frequency). Since Florey,[11,12] Smith,[13] and Horstmann[14,15] it has been postulated that the generation and distribution of lymphatic contractions depend exclusively on mechanical stimuli. Distension of the lymphatic wall activates the lymphatic contraction, which propels lymph to the next lymphatic segment. Later, in numerous studies performed both *in vivo* and *in vitro*,[3,8,9,16–23] it was shown that increases in transmural pressure caused positive inotropic and chronotropic effects in lymphatics. Lymphatics from different tissues and species attain maximal pumping at different values of intravascular pressure. A further increase of pressure inside the lymphatics causes overdistension of the lymphatic wall and diminishes pumping. Because of the importance of pressure stimuli for lymphatic contractility, the concept that distension stimuli are necessary to generate lymphatic contractions has dominated the literature for many decades.

However, beginning in the 1970s, several studies reported that lymphatic vessels could contract in a coordinated fashion without distension stimuli,[8,16,24,25] and that lymph flow could occur even when transmural pressure is zero cm H_2O.[18] Moreover, experiments performed on lymphatics from different tissues and species showed a high percentage of cases in which the contractile wave propagates in a retrograde direction along the vessel.[1,9,26–28] With the presence of highly competent valves in lymphatics, contractions of upstream lymphangions could only be activated after the contraction of a downstream lymphangion by the retrograde propagation of electrical excitation. At low or normal levels of lymph formation, upstream lymphangions are often nearly empty.[29] Thus, stretch-dependent activation of their contractions in such situations is very unlikely. Particularly, Mislin and Rathenow noted that the contractile wave could propagate in the retrograde direction through several lymphangions unrelated to the increase of the local transmural pressure.[1]

In recent studies,[30,31] we performed several series of experiments designed to evaluate the correlation between the presence of distension/stretch stimuli and the contractile activity of bovine and rat mesenteric lymphatics. In one set of experiments performed on isolated/cannulated bovine lymphangions, we created conditions of controlled, low-amplitude sinusoidal fluctuations of input pressure similar to those seen *in vivo*. In 80% of cases, we found little or no correlation of lymphangion contraction with pressure maxima.

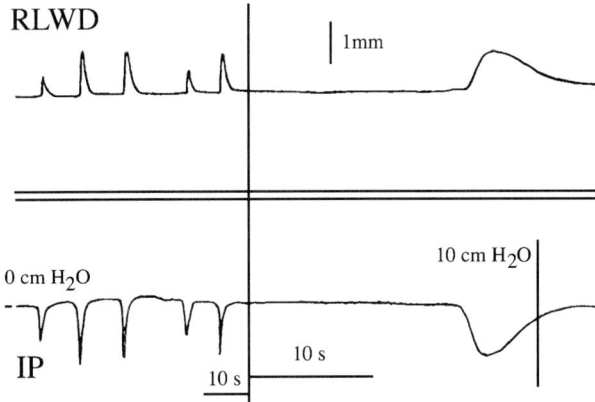

FIGURE 1. Portion of a recording from an isolated/cannulated bovine mesenteric lymphangion at zero cm H_2O of intravascular pressure. RLWD, radial lymphangion wall displacements (decrease of wall diameter [contraction] = ascending line); IP, intraluminal pressure (the mark at the beginning of line depicts a level of zero cm H_2O.) Note the change in chart speed during this recording. The input cannula was fixed; the output cannula was not fixed; and longitudinal tension was excluded. Input and output diastolic pressures were set to zero cm H_2O. (Redrawn from Gashev et al.[31])

In a separate set of experiments, we investigated the contractile behavior of both isolated bovine and rat mesenteric lymphatics at zero cm H_2O intraluminal pressure, and in the absence of radial and axial distension. After 40–45 minutes of exposure of specimens to such conditions, we observed stable (1–1.5 hours) spontaneous contractions of lymphatics. A fragment of recording from one such experiment with isolated bovine mesenteric lymphangions is shown in FIGURE 1. Moreover, during systole in bovine mesenteric lymphangions, the intraluminal pressure was periodically negative (–2 to –5 cm H_2O.) Hydrodynamic conditions similar to these could occur in lymphatic beds *in vivo* during periods of low levels of lymph formation. We proposed that the suction effect of contracting lymphatics under conditions of low filling could be an additional component of the active lymph pump that might be an important driving force for emptying the initial lymphatics.

These data lead to the reasonable conclusion that the distension of the lymphatic wall by intraluminal pressure is an important factor which modulates pacemaker activity in lymphatics, but is not a mandatory factor for pacemaking.

MODULATION OF THE ACTIVE LYMPH PUMP BY FLOW: REGIONAL DIFFERENCES IN LYMPHATIC CONTRACTILE BEHAVIOR

Lymph flow is the result of a complicated combination of passive and active driving forces. They vary regionally, so that lymph flow conditions are not the same in the different parts of body. In both the mesenteric lymphatics and the thoracic duct, the rate of lymph formation is a critical passive driving force. However, in the thoracic duct, the suction produced during inspiration and the low transient pressures in the central veins are also significant additional driving forces affecting lymph flow. Recently we conducted experiments[32,33] to test the hypothesis that different conditions in the thoracic duct and mesenteric lymphatics influence the active lymph pump. We investigated the effects of intraluminal pressure and imposed flow on the active lymph pumps in isolated rat mesenteric lymphatics and thoracic ducts. We found that mesenteric lymphatics have a greater ability to increase pumping during elevated transmural pressures. As shown in FIGURE 2, at similar pressure levels, the contraction amplitude is larger and contraction frequency is higher in the mesenteric lymphatics than in the thoracic duct. Thus, at the same pressure levels, the mesenteric lymphatics pump 3–4 times more fluid per vessel volume than does the thoracic duct. Mesenteric lymphatics also

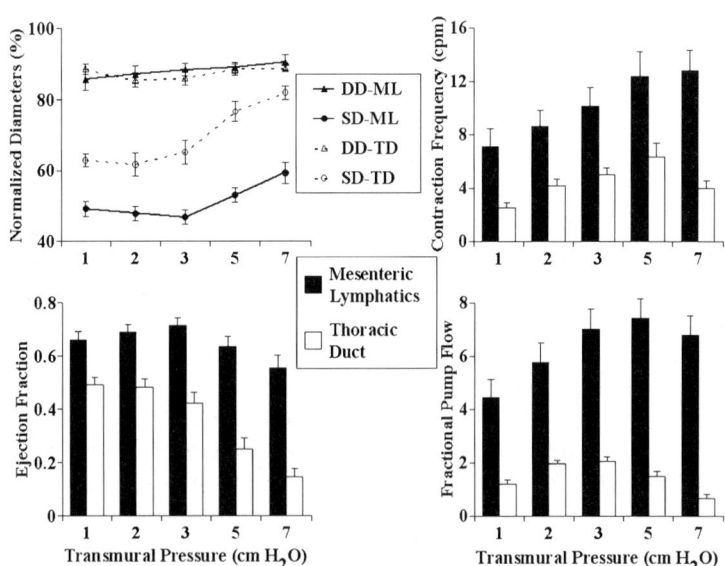

FIGURE 2. Comparison of the influences of transmural pressures on the active lymph pump of rat mesenteric lymphatics (ML; $n = 8$) and thoracic duct (TD; $n = 6$). DD, diastolic diameter; SD, systolic diameter.

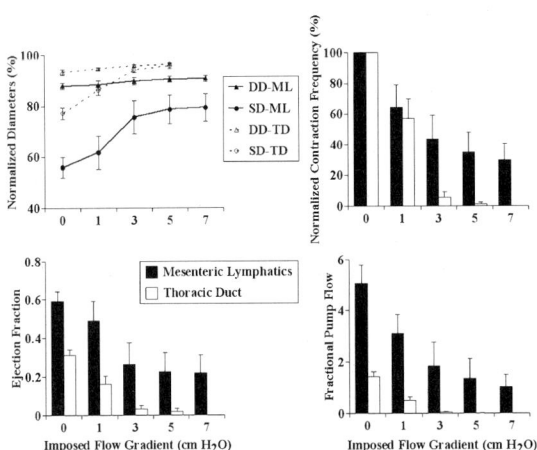

FIGURE 3. Comparison of the influences of imposed flow gradients on the active lymph pump of rat mesenteric lymphatics (ML; $n = 7$) and thoracic duct (TD; $n = 9$). DD, diastolic diameter; SD, systolic diameter.

pump more effectively at higher pressures. The maximum pumping in the thoracic duct occurs at 2–3 cm H_2O transmural pressure, while the maximum pumping in mesenteric lymphatics is seen at 5 cm H_2O.

We found that lymphatics from the different regions of body responded differently not only to the changes in intraluminal pressure, but also to the changes in imposed flow. We demonstrated an inhibition of the microlymphatic pump by controlled increases in an imposed axial lymph flow with constant transmural pressure. FIGURE 3 illustrates the comparative influence of the imposed flow on the active lymph pump in rat mesenteric and rat thoracic duct. The thoracic duct is more sensitive to imposed flow. The active pump in thoracic duct was essentially completely inhibited (97%) by an imposed flow of 3 cm H_2O, whereas the inhibition of the mesenteric lymph pump was significantly smaller (~60%). Mesenteric lymphatics even at the highest imposed flow still had an effective lymph pump.

As mentioned above, there are many forces that influence lymph flow. Lymphatics from the different regions of body have their own specific hydrodynamic conditions. For example, because of anatomic position and size, the outflow resistance of the mesenteric lymphatic bed is significantly higher than that in the thoracic duct. In view of such regional differences, one might also expect regional differences in the role of the active lymph pumps. As noted above, we found that responses to imposed flow were not identical for mesenteric and thoracic duct lymphatics. Even at higher imposed flows, the active pump in mesenteric lymphatics was comparatively more effective, whereas lower rates of imposed flow in the thoracic duct caused almost complete inhibition of the active pump. In other words, even at a high rate of flow generated by increased lymph formation, the mesenteric lymphatics may still need a strong active pump to overcome the high outflow resistance. In the thoracic duct under the same conditions, there is no need to develop additional forces by the active contractions of the duct, because passive driving forces such as the suction effect of inspiration and the influence of low or negative pressures in the central veins might suffice to drive lymph centripetally. Indeed, active pumping would only increase the outflow resistance of the contracting vessel.

Our general conclusion from these experiments is that the thoracic duct could be described as a principally conductive vessel whereas the mesenteric lymphatics are strong pumping vessels. Due to the complicated hydrodynamic conditions in lymphatic beds, the passive and active lymph pumps might at times work together to propel lymph centripetally. But in other cases, for example, under conditions of enhanced lymph flow, flow-mediated inhibition of the active lymph pump could serve to decrease lymphatic outflow resistance and save metabolic energy when the driving force of the passive lymph pump is enough to propel lymph. The rate of passive flow in lymphatics, as well as the value of intraluminal pressure, are important modulators of the active lymph pump, and flow-induced inhibition of contractile activity could be described as an autoregulatory mechanism of lymph pumping.

Our studies were designed to evaluate the substantial functional differences between the lymphatics from two different regions of the rat. Recently, it was shown, in isolated rat iliac microlymphatics,[34] that increased imposed flow decreases the amplitude and increases the frequency of contractions. In this study, the frequency response of lymphatics to flow was the opposite of what we have seen in both the rat mesenteric lymphatics and the thoracic duct. Since we have shown strong differences between the characteristics of the responses of active lymph pumps in rat mesenteric lymphatics and rat thoracic duct to pressure and flow, it is possible that there are more profound differences in the reactivity of lymphatics from the iliac and other regions. Such differences could reflect, not only differing reactivity of the lymphatics to changes in hydrodynamic conditions, but also regional differences in the activity of chemical regulatory pathways for the lymph pump.

CONTRACTILITY OF LYMPHATICS IN HUMANS

In spite of discoveries of the last decades, there is still no generally accepted model that adequately describes the mechanisms of lymph transport and its regulation. Moreover, some of the results obtained from different species are inconsistent with one another. Most of the recent findings that concern the generation and regulation of lymph flow were obtained from animal experiments, but some scientists and physicians still have doubts about the presence and importance of intrinsic contractility of human lymphatics. In general, we don't have enough human lymphatic data to verify the extrapolations of animal data to lymph flow in humans. Mechanistic evaluations of the regulation of lymph flow in humans are critical to ongoing attempts to discover the pathogenesis of and effective treatment for lymphatic diseases like lymphedema. Our knowledge of the generation and regulation of human lymph flow is very limited and consists primarily of visual observations of the contractile activity during clinical manipulations within a few mechanistically designed experiments (see Ref. 35 for review). Currently this is reasonable evidence of the presence of spontaneous contractile activity in the human lymphatic vessels. But the mechanisms that regulate lymph flow in humans are still not well understood. We propose that the active lymph pumps in humans may have greater regional differences in contractile function than has been seen in animals because the upright posture for bipedal humans creates an additional outflow resistance for most lymphatics in the lower part of the body. Even the thoracic duct, where highly developed muscular layers were found in humans,[36] requires a strong active lymph pump to propel lymph against the gravitational forces encountered in the upright human. However, the position of the lymphatics above the level of the juncture of the right lymphatic duct with the great veins provides conditions more favorable for lymph flow because of the additional action of gravitational forces. The inhibitory effect of flow in lymphatics could potentially serve two purposes here: to decrease outflow resistance and to conserve energy in the regions where passive forces are enough to move lymph. Thus, potentially, the lymphatics of the upper part of human body could be more sensitive to flow/shear.

Moreover, there is great variability in the forces and factors that generate lymph flow that depends upon the varying activities of the human body during the day. Nevertheless, there is very limited comprehension of mechanisms that modulate the active lymph pump in humans. Such investigations assume great importance in the ongoing attempts to uncover the pathogenesis and effective therapies for lymphedema. There are millions of people in the United States and hundreds of millions people worldwide affected with this disease. Investigators who undertake the study of lymphatic dysfunction and ways to treat it do so in a situation where the mechanisms responsible for normal lymph flow in humans are still unclear.

REFERENCES

1. MISLIN, H. & D. RATHENOW. 1962. Experimentelle Untersuchungen ueber die bewegungskoordination der Lymphangione [German]. Rev. Suisse Zool. **69**: 334–344.

2. MISLIN, H. 1966. Structural and functional relations of the mesenteric lymph vessels. *In* New Trends in Basic Lymphology; Proceedings of a Symposium held at Charleroi (Belgium), July 11–13, 1966. Experientia Suppl. **14:** 87–96.
3. OHHASHI, T., T. AZUMA & M. SAKAGUCHI. 1980. Active and passive mechanical characteristics of bovine mesenteric lymphatics. Am. J. Physiol. **239:** H88–95.
4. ORLOV, R. S. & G. I. LOBOV. 1984. Ionic mechanisms of the electrical activity of the smooth-muscle cells of the lymphatic vessels [Russian]. Fiziol. Zh. SSSR Im. I. M. Sechenova **70:** 712–721.
5. ALLEN, J.M. & N.G. MCHALE. 1986. Neuromuscular transmission in bovine mesenteric lymphatics. Microvasc. Res. **31:** 77–83.
6. VAN HELDEN, D.F. 1993. Pacemaker potentials in lymphatic smooth muscle of the guinea-pig mesentery. J. Physiol. **471:** 465–479.
7. VON DER WEID, P.Y., M.J. CROWE & D.F. VAN HELDEN. 1996. Endothelium-dependent modulation of pacemaking in lymphatic vessels of the guinea-pig mesentery. J. Physiol. **493:** 563–575.
8. MCHALE, N.G. & I.C. RODDIE. 1976. The effect of transmural pressure on pumping activity in isolated bovine lymphatic vessels. J. Physiol. **261:** 255–269.
9. BENOIT, J.N., D.C. ZAWIEJA, A.H. GOODMAN, *et al.* 1989. Characterization of intact mesenteric lymphatic pump and its responsiveness to acute edemagenic stress. Am. J. Physiol. **257:** H2059–2069.
10. GASHEV, A. A. 1991. The mechanism of the formation of a reverse fluid filling in the lymphangions [Russian]. Fiziol. Zh. SSSR Im. I .M. Sechenova **77:** 63–69.
11. FLOREY, H. 1927. Observations on the contractility of lacteals: Part I. J. Physiol. **62:** 267–272.
12. FLOREY, H. 1927. Observations on the contractility of lacteals: Part II. J. Physiol. **63:** 1–18.
13. SMITH, R. 1949. Lymphatic contractility: a possible intrinsic mechanism of lymphatic vessels for the transport of lymph. J. Exp. Med. **90:** 497–509.
14. HORSTMANN, E. 1952. Über die funktinelle Struktur der mesenterialen Lymphgefasse [German]. Morphol. Jahrb. **91:** 483–510.
15. HORSTMANN, E. 1959. Beobachtungen zur Motorik der Lymphgefasse [German]. Pflugers Arch. **269:** 511–519.
16. MCHALE, N.G. & I.C. RODDIE. 1975. Pumping activity in isolated segments of bovine mesenteric lymphatics. J. Physiol. **244:** 70P–72P.
17. ORLOV, R.S. & T.A. LOBACHEVA. 1977. Intravascular pressure and spontaneous lymph vessels contractions [Russian]. Bull. Exp. Biol. Med. **83:** 392–394.
18. REDDY, N.P. & N.C. STAUB. 1981. Intrinsic propulsive activity of thoracic duct perfused in anesthetized dogs. Microvasc. Res. **21:** 183–192.
19. ORLOV, R.S. & G.I. LOBOV. 1984. Mechanism of action of intravascular pressure on the electrical and contractile activity of lymphangions [Russian]. Fiziol. Zh. SSSR Im. I. M. Sechenova **70:** 1636–1644.
20. HOGAN, R.D. & J.L. UNTHANK. 1986. Mechanical control of initial lymphatic contractile behavior in bat's wing. Am. J. Physiol. **251:** H357–H363.
21. HAYASHI, A., M.G. JOHNSTON, W. NELSON, *et al.* 1987. Increased intrinsic pumping of intestinal lymphatics following hemorrhage in anesthetized sheep. Circ. Res. **60:** 265–272.

22. GASHEV, A.A. 1989. The pump function of the lymphangion and the effect on it of different hydrostatic conditions [Russian]. Fiziol. Zh. SSSR Im. I.M. Sechenova **75:** 1737–1743.
23. EISENHOFFER, J., S. LEE & M.G. JOHNSTON. 1994. Pressure-flow relationships in isolated sheep prenodal lymphatic vessels. Am. J. Physiol. **267:** H938–943.
24. MAWHINNEY, H.J. & I.C. RODDIE. 1973. Spontaneous activity in isolated bovine mesenteric lymphatics. J. Physiol. **229:** 339–348.
25. HARGENS, A.R. & B.W. ZWEIFACH. 1977. Contractile stimuli in collecting lymph vessels. Am. J. Physiol. **233:** H57–65.
26. MCHALE, N.G. & M.K. MEHARG. 1992. Co-ordination of pumping in isolated bovine lymphatic vessels. J. Physiol. **450:** 503–512.
27. ZAWIEJA, D.C., K.L. DAVIS, R. SCHUSTER, et al. 1993. Distribution, propagation, and coordination of contractile activity in lymphatics. Am. J. Physiol. **264:** H1283–1291.
28. CROWE, M.J., P.Y. VON DER WEID, J.A. BROCK, et al. 1997. Co-ordination of contractile activity in guinea-pig mesenteric lymphatics. J. Physiol. **500:** 235–244.
29. GASHEV, A.A., R.S. ORLOV, A.V. BORISOV, et al. 1990. The mechanisms of lymphangion interaction in the process of the lymph movement [Russian]. Fiziol. Zh. SSSR Im. I.M. Sechenova **76:** 1489–1508.
30. GASHEV, A.A. & D.C. ZAWIEJA. 1999. Lymphatic contractions: the role of distension mechanisms. FASEB J. **13:** A11.
31. GASHEV, A.A., R.S. ORLOV & D.C. ZAWIEJA. 2001. Contractions of the lymphangion under low filling conditions and in absence of distension stimuli: a possibility of the suction effect [Russian]. Ross. Fiziol. Zh. Im. I.M. Sechenova **87:** 97–109.
32. GASHEV, A.A. & D.C. ZAWIEJA. 2001. Comparison of the active lymph pumps of the rat thoracic duct and mesenteric lymphatics. *In* 7th World Congress for Microcirculation, Sydney, Australia. :P1–19.
33. GASHEV, A.A., M.J. DAVIS & D.C. ZAWIEJA. 2002. Inhibition of the active lymph pump by flow in rat mesenteric lymphatics and thoracic duct. J. Physiol. **540:** 1023–1037.
34. KOLLER, A., R. MIZUNO & G. KALEY. 1999. Flow reduces the amplitude and increases the frequency of lymphatic vasomotion: role of endothelial prostanoids. Am. J. Physiol. **277:** R1683–1689.
35. GASHEV, A.A. & D.C. ZAWIEJA. 2001. Physiology of human lymphatic contractility: a historical perspective. Lymphology **34:** 124–134.
36. ORLOV, R.S., A.V. BORISOV & R.P. BORISOVA. 1983. Lymphatic Vessels: Structure and Mechanisms of Contractile Activity [Russian]. Nauka. Leningrad.

Part 4: Aspects of Lymphatic Biology and Disease

Panel Discussion

The Role of Bacterial Infection in Filarial Lymphedema

WALDEMAR OLSZEWSKI (*Polish Academy of Sciences*): I challenge the commonly accepted notion that filarial lymphedema is caused by the parasite. It might be that the primary insult to the lymphatic system is caused by the parasite, but that may not necessarily be the main factor. In my opinion, it is the secondary, bacterial infections that create the major damage in the lymphatics and the nodes.

We have performed on the order of 300 inguinal lymph node biopsies and were unable to find any microfilaria or any debris of the adult forms in our preparations, nor was there evidence for any cellular reactions against an organism in the nodes. Furthermore, we could not detect any increased level of anti-parasitic antibody; the immunoglobulin concentration in the lymph was approximately 25% of what was present in the plasma, indicating that, even if there were an increase in this fraction of globulin, the globulin would have been produced systemically, and not locally. In 1997, we published in the *American Journal of Tropical Medicine* that it is largely saprophytic *Staphylococcus* that is most often responsible for the changes. I believe that microbes that penetrate the skin and are physiologically transported linearly along the lymphatics to the nodes are responsible for what we see in lymphedematous tissue.

As indirect clinical evidence for this hypothesis, I can attest to the fact that, when we give even a low dose of penicillin, all of the clinical manifestations of dermatolymphadenitis disappear within hours. Consequently, we have proposed this protocol of long-term treatment or prophylaxis with penicillin to the World Health Organization (WHO). Thus far, I believe that more than one or two million people in India are receiving these injections and they are free of the attacks of dermatolymphadenitis, of recurrence of the so-called filarial fever.

PATRICK LAMMIE (*Centers for Disease Control and Prevention*): It is clear that the final stages of the disease process are dominated by bacterial infections, and I wouldn't argue that bacterial infection is important in the early stages, either. Nevertheless, in many of the situations where I work, in the

Americas and in the South Pacific, where the transmission of filariasis has effectively been eliminated, lymphedema and elephantiasis disappear as public health problems. So, in that context, it is clear that the worm is providing the initial insult.

In terms of our immunological data, the hypothesis that we would like to test in the field is that there is something unusual about the nature of the inflammatory response in these patients. I believe that this inflammatory response both drives the disease process, in terms of the response to the bacterial superinfections, and also provides the patients with a measure of protection against repeated infections. Again, if I were to present a comparison of the prevalence of infection between my lymphedema patients in Haiti and age-matched controls, the statistical significance of the differences would be astounding.

The Effect of Inflammation on Lymphatic Contractility

ANATOLIY GASHEV (*Texas A&M University*): It has recently been published that lymphatic contractility, measured indirectly, is inhibited in people that have lymphedema. It is also known that inflammatory products, including oxygen radicals, nitric oxide, and other agents that would theoretically be released in the immune response, can actually serve as potent inhibitors of lymphatic contractility.

LAMMIE: If we consider the capacity of the worm to cause lymphangiectasia, then we would expect to find compromised lymphatic function in 70% of the Haitian population. Yet the published studies performed in other settings would suggest that active infection, if it has any effect, is associated with relatively normal lymphatic flow.

Therefore, the unanswered question would be: to what extent does the parasite actually promote development of a collateral lymphatic circulation that can compensate for the loss of function in the structures where the worm actually resides? There is some published evidence that, with the death of the worm, the collaterals disappear.

GASHEV: Yes, but is it the fact that you have a very enhanced immune response, with associated inflammation, that could also be leading to the gross development of lymphedema?

LAMMIE: Yes, and that may be testable.

Adipogenesis or Fat Accumulation?

TIM PADERA (*Massachusetts Institute of Technology*): In the fat accumulation of lymphedema, is it cellular proliferation, or is it simple incorporation of fat into those cells?

EVAN ROSEN (*Beth Israel Deaconess Medical Center*): More than 15 years ago, studies were performed on rats fed a high-fat diet. Adipogenesis in the fat pads was documented by ^{3}H-thymidine incorporation. Although the comparable experiment cannot be performed in humans, one can isolate and differentiate pre-adipocytes from a human fat pad. Such experiments have been performed with specimens from individuals up to 80 years of age.

There is certainly no doubt that the capacity to proliferate and to differentiate in the human fat pad is present at almost any age tested. However, the amount of fat accumulation that is seen in the lymphedematous limbs is far too much to be accounted for by simple hypertrophy of pre-existing fat cells alone. There simply aren't sufficient numbers of pre-existing fat cells in the limb to account for the patterns that are seen. I think that, almost certainly, there must be recruitment and proliferation of new fat cells in those tissues. There is a finite limit to how large a fat cell can become, but there is no theoretical limit to how much fat mass a human can accumulate.

PADERA: What is the initial stimulus for the accumulation of adiposity?

ROSEN: The answer to this question is not known, but one can speculate. In one published experiment, rabbit pre-adipocytes were stimulated to differentiate in culture in the presence of rabbit lymph. It was demonstrated to be the chylomicron fraction in the lymph that was the inducing factor. So perhaps it is either the presence of the lipid itself or, more likely, a lipid-soluble ligand for PPRγ or some comparable substance that triggers adipogenesis.

Subcellular Localization of PROX-1

CARLA MOUTA (*Maine Medical Center*): Dr. Oliver, I am very interested in your work on PROX-1 and how this molecule can affect such diverse cell populations as neurons, hepatocytes, and hepatic endothelial cells.

One of the ways to regulate PROX1 function, at least in *Drosophila*, is through the regulation of its subcellular localization from the cytoplasm to the nucleus. It has been shown that, even in mammalian cells, there are ways to control the subcellular localization of PROX and it has been mapped to specific mutants.

Have you been able to create transgenic models in which you overexpress PROX mutants that have different subcellular localization abilities?

GUILLERMO OLIVER (*St. Jude Children's Research Hospital*): PROX was cloned through its similarity to the fly properagene. In *Drosophila*, PROSPER is very important in the central nervous system because it exhibits as a metric distribution upon division. Asymmetric distribution determines the fate of the ganglia mother cell or the neuroblast.

The similarities between PROX and PROSPER at the sequence level are only in the HOMEO and the PROS domain and, at the functional level, are very limited. Indeed, as you mentioned, PROX-1 doesn't have the domain required in the fly for asymmetric distribution. Also, that domain, in the fly, is crucial for interaction with other factors, like Miranda, Stelfin, and Scitaval; this does not occur in the mammalian counterpart.

It has been shown through staining that, in certain cells in the lens, PROX-1 is in the cytoplasm. This is not to say that it is asymmetrically distributed, but it is present both in the cytoplasm and in the nuclei. What functional role this plays is not known. However, I must say that, in our hands, using β-gal antibody staining, there has never been identification of PROX in the cytoplasm. We clearly see differing concentrations of PROX protein in different cell types and, on the basis of our having sufficient phenotype, we are sure that that is extremely important.

CARLA MOUTA (*Maine Medical Center Research Institute*): I think then that it will be very important to compare reagents because I, myself, have seen a very interesting shift in PROX distribution in the human liver between cytoplasm, exclusively, and nucleus, exclusively, as an example, in adipocytes. So, even in the human, there is evidence that the protein can accumulate in one place or another, which should be expected to control expansion.

OLIVER: Yes. However, what I'm trying to stress is that the importance of PROSPER in the fly is not the cytoplasmic distribution but the asymmetric distribution. It goes to the cortical layer in the cytoplasm and, when the cell divides, the localization of PROSPER is what makes the difference. That attribute is the one that has not, thus far, been seen in the mammalian counterpart. I'm not disputing the cytoplasmic or nuclear localization. We just haven't seen it thus far in our studies. I don't know how functionally relevant it may be.

Cell Surface Receptors for Filarial Parasites in Lymphatic Endothelia?

MOUTA: Dr. Lammie, has anyone examined the interaction of these parasites, and the bacteria that live within them, with lymphatic endothelial cells in culture? Is there any evidence that the lymphatic endothelial cells contain specific cell surface receptors for these molecules that are absent from the blood vessel endothelial cells?

LAMMIE: The simple answer is no, it's not been done. The worms are difficult to culture, as you might expect, given the host specificity. It would be wonderful to develop such a model system.

The Role of PROX-1 in Adult Lymphatic Maintenance

HELEN HAYES (*Texas A&M University*): Dr. Oliver, do you have any evidence or hypothesis that PROX-1 is involved in any stage of adult lymphatic maintenance? Or do you think the absence of PROX-1 would have any effects in the adult stage? I heard you mention that you're doing a conditional knockout. Do you have any preliminary data?

OLIVER: At the moment, all that I can say is that we don't know. We know that PROX-1 is expressed in adult tissues. At the moment we are examining our heterozygous mice to see what happens with PROX in wound healing or inflammation. I think that this might enhance our understanding of lymphatic development and the role of the lymphatic response under these conditions.

The Prevalence of Lymphedema in the United States

SIMON SIMONIAN (*Georgetown University*): Dr. Lammie, what is the prevalence of non-filarial lymphedema in the United States?

LAMMIE: I don't have any idea of the prevalence, but it would be great to get a good estimate of the national prevalence.

Genetic Factors in Obesity and Adipogenesis

SIMONIAN: Dr. Rosen, why do some animals or humans have more fat accumulation or more adipogenesis than others?

ROSEN: Adipose mass reflects many factors. Certainly food intake is an important one. Energy expenditure is another. These are all under genetic control. In fact, all of the monogenic disorders that lead to obesity that have been described to this point have been outside the adipocyte. They have all been in the brain or in the pancreas or in other loci.

So, is obesity a condition of adipogenesis or adipocyte hypertrophy? Certainly, there is a hypertrophic component in most forms of common obesity. This is most directly related to excess energy intake and probably is genetically regulated.

Variability in Lymphangion Contractility

SIMONIAN: Dr. Gashev, I am interested in the fact that the enteric lymphangion contracts more than, for example, the thoracic lymphangion. What is the lymphangion doing in the leg, that is, at the most dependent position in the pumping effect? Is its contraction more powerful than a lymphangion situated higher in the chain?

GASHEV: Let me refer, for example, to a publication of Dr. Olszewski. He showed very high release of pressure and very high conductivity of the human leg lymphatics. So, yes, there is greater contractility in these vessels than in their mesenteric counterparts.

ALBERT MILLER (*Northwestern University*): I would like to comment about the flow of lymph in the heart, which is unique. When the ventricular muscle contracts, it pushes the lymph from endocardium to epicardium.

Then, when the heart dilates, the collecting vessels on the surface of the heart push the lymph into the main collecting vessels that ascend directly to the right lymphatic duct. When we cannulate the lymphatic before it enters the right lymphatic duct, near the cardiac lymph node, the flow is continuous. It is not irregular. We think the flow is continuous because there are numerous valves along the entire surface.

Also, we have found that, when cannulating a major lymphatic, if we are rough with it, it markedly constricts. We must wait, sometimes even with application of papaverine, to induce relaxation so that we can cannulate. This is a unique flow phenomenon that is relevant to the discussion.

The Relationship of PROX-1 Heterozygosity to Human Disease

MILLER: Dr. Oliver, there is a recognized, rare disease entity of infants who are born with profound generalized ectasia of the lymphatics. Effectively, the lymphatics are obstructed and these children die at a very young age. Is there any relationship of this entity to your subject matter?

OLIVER: What I have presented here is documented in our heterozygous mice. We know that the pattern of the lymphatic vessels is in some manner affected. It is likely that the effect is dose-dependent. In the homozygous mice, you remove PROX and, basically, there are no lymphatics. So, of course, we have been interested to discover whether there is any correlation in the literature for a pathological lymphatic alteration that might be related to PROX.

Of course, mutations in PROX will be lethal; therefore, the only possibilities would be in heterozygous conditions—perhaps a condition that would mimic what we see in our mice.

ROBERT FERRELL (*University of Pittsburgh*): Dr. Oliver, we were optimistic that at least some of the families that have primary lymphedema would harbor mutations in PROX-1 and we were disappointed, I suppose, by our failure to find any point mutations in PROX-1. Given your work, would you speculate that haplo-insufficiency for PROX-1 might be lethal in humans?

OLIVER: I don't know. In our experience with five or six different genetic backgrounds in mice, it was very difficult and time-consuming to find one in which there was some survival of the heterozygous PROX mice. In any other background, the heterozygous states were lethal. However, our experience indicates that there are many differences between mice and humans. Therefore, this is a question that I cannot address.

The Role of Wolbachia in Disease Manifestation

BREN GANNON (*Flinders University*): Dr. Lammie, was the assumption that, for the *Wolbachia* to have any effect, it required the death of the worm?

I'm wondering whether there is any possibility that the *Wolbachia* is either partly shed as a result of an immune response against the worm or whether, in fact, *Wolbachia* might be releasing some toxin that might be part of the system that is driving the response of the patient.

LAMMIE: There is no evidence that the *Wolbachia* are released except upon the death of either the adult or microfilarial stages for the worm. So, at this point, we don't have any other markers that would allow us detect a local release of a toxin or other product. My assumption, actually, is no, because there is very little evidence of an inflammatory response around a living worm. The inflammatory response is sparked by the death of the parasite, independent of the stage.

Fat Accumulation in Lymphedema: Considerations from Comparative Physiology

GANNON: Dr. Rosen, in lymphedema, the development of fat seems to be associated with the initial presence of large amounts of water and salts in the fluid in the limb. I wonder whether the body is trying, as it were, to compartmentalize the water. I am thinking of the lesson from comparative physiology where, in the camel, the existence of the storage of water occurs as fat. In fact, what they are storing is hydrogen. They derive oxygen from the atmosphere, when they need it, for an extra supply of water; storing the water as fat overcomes many osmotic and hydraulic problems.

ROSEN: I'm impressed that you realize that camel humps are fat and not water. But no, I cannot think of a bioenergetic reason why that would occur.

GANNON: One of the reasons is that fat is 30% lighter than water.

ROSEN: That's an interesting idea.

Cell Lineage Analysis of PROX-1

JOERG WILTING (*University of Göttingen*): Dr. Oliver, I completely agree that PROX-1 is very important for lymphatic development, and your results clearly show that. And it is obviously a very good marker for lymphatic endothelial cells. We have completed some studies in chick embryos where we have studied PROX-1. We also have a manuscript in press that also shows that PROX-1 is a very good marker in embryonic and adult tissues in the human.

However, I don't agree that your data support the theory of Sabin. All that you have observed is PROX-1-positive cells. We have performed grafting experiments that demonstrate that even before PROX-1 is expressed, there is a lymphangiogenic potential in these PROX-1-negative tissues. So, having seen PROX-1 positive cells, you cannot actually determine whether these cells, which might have been PROX-1-negative, integrate into the vein. You cannot see whether the cells go out or whether they go in. This is something we can further discuss in the workshop. I think that additional cell lineage studies are needed.

OLIVER: Yes; I have no doubt that it will be great to perform some cell lineage analysis. However, I would like to respond to your question. First of all, within the transgenic mice that I showed, the only place where they express PROX-1 is in the endothelial cells. One observes premature and random budding from the vein, so there are cells that are emerging from the vein. Therefore, we are not targeting a lymphoangioblast or other similar cell in these mice. We are targeting endothelial cells with PROX-1 expression. For me, this argues that the lymphatic endothelial cells that are originating in the transgenic mice are arising from endothelial precursors.

WILTING: But are you targeting cells that you think are the precursors?

OLIVER: I'm targeting endothelial cells.

WILTING: Perhaps if you target other cells, PROX-1 may be so strong that you even drive any other cell type into a lymphatic phenotype.

OLIVER: We are using an endothelial-specific promoter.

Target Genes for PROX-1

QUESTION: Dr. Oliver, do you have any evidence of what is downstream to PROX-1?

OLIVER: Finding target genes is not easy. For PROX, not much is known. PROX plays a major role in a variety of tissues. In many respects, its role seem to be similar, like control of the cell cycle, control of some migration, and control of cellular differentiation. We have concluded, despite the differences among these tissues, perhaps the mechanism through which PROX exerts its functional role may be conserved. Therefore, we have performed microarray analysis in all these tissues, mutant and wild-type, at different stages. The idea is to focus on those genes that appear in common among all these different tissues. To our surprise, we have found approximately 20 or 30 genes that correspond in all tissues. We are targeting these genes at the moment to see whether, indeed, they may be downstream targets of PROX, directly or indirectly.

PROX probably controls a very important and unique step in development. In general, when a cell exits the cell cycle, it migrates and differentiates. I think, in every tissue, PROX-1 in some manner controls or influences these important steps.

PROX-1 Expression in the Central Nervous System

QUESTION: Dr. Oliver, is PROX-1 expressed in the brain of these fetuses?

OLIVER: Yes, PROX-1 is expressed all along the central nervous system. It is also expressed in the brain. The expression of PROX follows the process of neurogenesis. You can find it, for example, in the basal ganglia early, on about E14.5 in the ventral thalamus, and later in the dorsal thalamus. It is expressed in the cerebellum and in the hippocampus.

QUESTION: Do you know whether it is eventually down-regulated as a lymphatic marker?

OLIVER: In the spinal cord, yes. It effectively follows the process of neurogenesis in the spinal cord. In the brain it does not, so that, in adult cerebellum and hippocampus, it is still expressed. What it is doing there, we don't know.

Anti-inflammatory Therapy for the Anti-filariatic Response

QUESTION: Dr. Lammie, have anti-inflammatory agents been identified that prevent the attack of inflammatory cells against the filarial organism?

LAMMIE: There are a lot of hypotheses, but no proof.

The Role of Interstitial Stress in Lymphatic Function and Lymphangiogenesis

MELODY A. SWARTZ AND KENDRICK C. BOARDMAN, JR.

Department of Biomedical Engineering, Northwestern University, Evanston, Illinois 60208, USA

ABSTRACT: The management and control of tissue fluid balance depends on the highly regulated orchestration of various interstitial factors. In particular, lymphatic function, lymphatic biology, and development (lymphangiogenesis), and the extracellular matrix all contribute to interstitial fluid balance. In light of the dynamic interdependence of these factors, our lab has been working towards establishing a mechanical-molecular picture of the process of lymphangiogenesis—that is, bridging the physiological context of lymphangiogenesis with its molecular regulation by studying the coordination of mechanical forces, ECM development, lymphatic biology, and lymphatic capillary organization and development. Our working hypothesis is that the physiological driving force for lymphangiogenesis is the need for organized interstitial fluid flow. This paper will outline the rationale and background for such an approach and highlight some of the recently published findings of our lab and others that support this concept.

KEYWORDS: interstitial stress; lymphatic function; lymphangiogenesis; growth factors

INTRODUCTION

The process of nourishing cells and removing waste products involves extravasation of fluid, solutes, and proteins from the blood, diffusion of solutes and convection of proteins through the interstitial space, and reabsorption into the postcapillary venules or uptake by the lymphatics. Thus, the interstitial space and the lymphatic system are critical components of the microcirculation because they maintain the driving forces necessary for convection

---

Address for correspondence: Prof. Melody A. Swartz, Department of Biomedical Engineering, Northwestern University, 2145 N. Sheridan Road, Evanston, IL 60208-3107. Voice: (847) 467-6668; fax (847) 491-4928.
m-swartz2@northwestern.edu

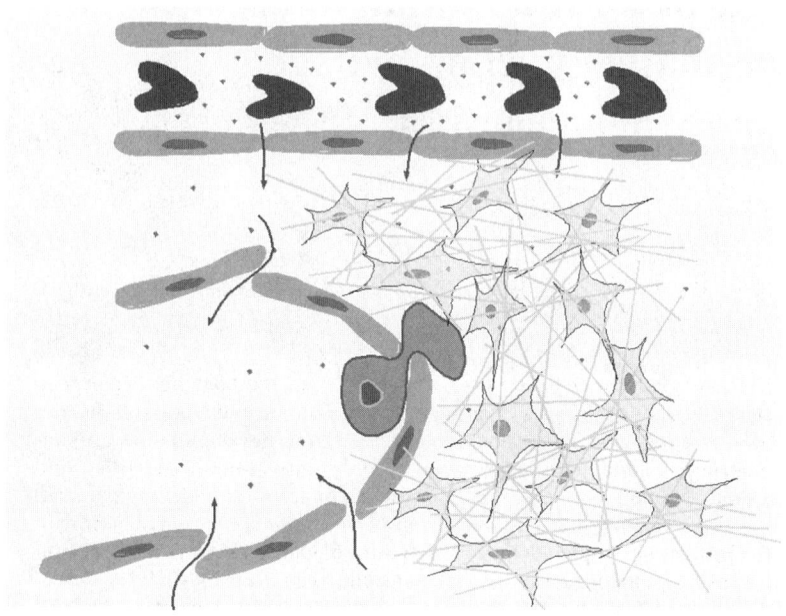

FIGURE 1. The microcirculation consists of the blood capillaries, lymphatic capillaries, and interstitial space containing ECM. These three components are mechanically coupled.

and tightly regulate interstitial fluid balance (FIG. 1). After uptake by the initial lymphatics (i.e., lymphatic capillaries), lymph drains into the collect vessels and passes through several clusters of lymph nodes before returning to the bloodstream, completing the circuit of fluid transport. Because of these features, the lymphatic system serves as a major transport route for immune cells and disseminating tumor cells in addition to its primary function of providing interstitial fluid balance and macromolecular transport. Cells and particles experience lower flow rates and smaller shear stresses than they would in the blood circulation, and, moreover, the lymph nodes provide an environment in which immune cells reside and proliferate, and disseminating tumor cells take root and form metastatic tumors. Clearly, the roles of the lymphatics in tissue fluid balance, protein transport, and immunology are critical.

We have suggested that the mechanical environment of the interstitium plays a key role in the molecular regulation of lymphatic development. Despite many similarities at the cellular level, lymphatic and blood endothelial cells serve quite different functions, and the organizing principles of angiogenesis and lymphangiogenesis are likely to be related to these functions. While vascular capillaries provide nourishment to cells and their growth is governed by nutrient diffusion and consumption rates, the lymphatics play a mechanical role—in fact, they must act as a mechanical extension of the interstitium to be functional. We have proposed that the tight coupling of the mechanical and the molecular is a defining characteristic of lymphangiogenesis, specifically that certain mechanical stimuli are critical in its initiation and organization.

MECHANICS OF THE INTERSTITIAL–LYMPHATIC INTERFACE

Lymph formation occurs when fluid moves from the interstitium into the initial lymphatics. It is driven by local interstitial mechanical stresses, namely, interstitial fluid pressure and strain of the extracellular matrix (ECM), both of which can be affected by skeletal motion and massage as well as the slight strains associated with arterial pressure pulsations and vasomotion of neighboring arterioles. Lymph formation is mechanically coupled to lymph propulsion, that is, the transport of lymph from the initial capillaries to the larger vessels and eventually back to the blood. If there is blockage in the systemic route (e.g., removal of a lymph node), interstitial fluid may enter the initial lymphatics, but will eventually fill, back up, and cause edema. Likewise, if the interstitial–lymphatic interface is destroyed and lymphatic capillaries cannot function, interstitial fluid will not be drained from that local region despite the baseline systemic drainage forces.

With few exceptions, lymphatic capillaries do not have contractile capability and contain overlapping cell–cell junctions that give rise to intercellular

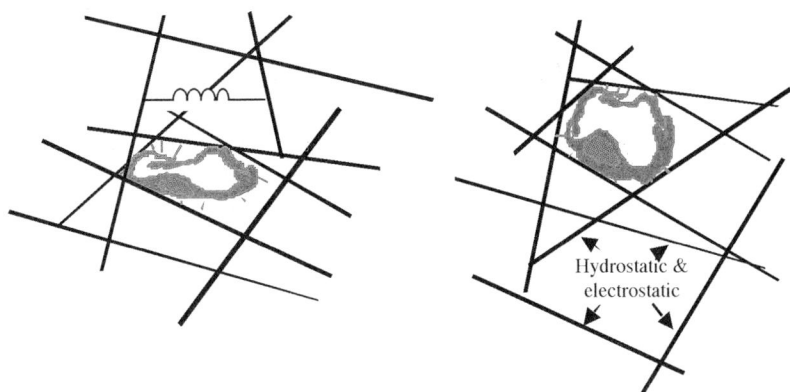

FIGURE 2. Schematic of the interstitial–lymphatic interface during lymph formation. *Left*: the capillary is mostly collapsed under resting conditions. *Right*: tissue strain (due to an interstitial stress like fluid accumulation) increases the luminal volume and also pulls open the intercellular junctions to facilitate lymph formation.

clefts, resulting in little to no exclusion of interstitial proteins during lymph formation.[1,2] To function, they rely on sensitive connections to the ECM called anchoring filaments, which tether the basal lamina to adjacent collagen fibers.[3,4] Thus the lymphatic capillary is highly attuned to interstitial stresses via radial tension by anchoring filaments to increase its luminal volume and open its intercellular clefts (FIG. 2). This creates a "tissue pump," or a small oscillating pressure gradient to facilitate lymph formation, while the overlapping cell–cell junctions constitute a "second valve system" to prevent retrograde flow from the lymphatic back into the interstitium.[5,6]

The mechanical integrity of the ECM, therefore, plays an integral role in lymphatic function, both because it facilitates lymph formation and also because its composition and architecture determine the elasticity, hydraulic conductivity, and hydration of a tissue.[7–10] While collagen fibers give structural stability and form the skeletal framework onto which tissue is composed, glycosaminoglycans (GAGs) are highly negatively charged macromolecules that make up a gel-like phase with water and interstitial proteins. In skin, the organization of collagen fibers and high GAG content govern its mechanical properties.[7,11,12] Furthermore, extensive and chronic degradation of the ECM eventually renders lymphatic vessels nonresponsive to the changes in the interstitium and therefore causes dysfunction.[13] Because of the critical importance of the ECM in tissue-fluid balance, the initial lymphatics need to be considered not as a separate and distinct entity from the surrounding interstitium, but rather as a functional extension of the interstitium itself. Indeed, this functional interaction suggests that the composition and architecture of the ECM are likely to play an important role in lymphangiogenesis.

LYMPHATIC GROWTH FACTORS

Although the lymphatic system was first described by Gasparo Aselli as early as the year 1627, research on the physiology and developmental biology of the lymphatic system lags far behind that of the vascular system. Budge[14] and Sala[15] were the first to describe the developing lymphatic system of chick embryos at the end of the 19th century. Recently, the field has attracted much attention in the angiogenesis community as the role of lymphangiogenesis has been implicated in tumor metastasis.[16–20]

VEGF-C, a member of the VEGF family of growth factors,[21,22] was the first growth factor demonstrated to specifically stimulate lymphatic endothelial cell (LEC) proliferation and mitogenesis.[23,24] In only a few years since its discovery, great strides have been made in identifying and characterizing this important molecule in lymphangiogenesis and metastasis. For example, we now know that the specific effects of VEGF-C on LECs depend on its proteolytic processing: the mature form of human VEGF-C stimulates both VEGFR-2 and VEGFR-3, and can therefore stimulate both angiogenesis and lymphangiogenesis, whereas the partially processed form preferentially binds and activates VEGFR-3[25] and has been shown to specifically stimulate LEC growth in tumors.[19] Structurally, VEGF-C is closely related to vascular endothelial growth factor-D (VEGF-D), which also binds to and activates VEGFR-2 and VEGFR-3 in a similar manner[26] and has been shown to stimulate both tumor angiogenesis and lymphangiogenesis.[27]

While VEGF-C and VEGF-D are now widely recognized as lymphangiogenic factors, their physiological (upstream) regulation has not yet been clarified. To illustrate this point, consider angiogenic growth factors: it is known that one of the most fundamental physiological regulator of angiogenesis is hypoxia, which in turn leads to the molecular signaling cascade that drives angiogenesis. This is in line with the most fundamental function of the blood circulation, tissue oxygenation. Such an analogy, relating physiological function to biological regulation, has not been elucidated in the process of lymphangiogenesis. With the rationale outlined above along with previous and ongoing studies, we attempt to show the extent to which the *physical environment* and *mechanical function* of lymphatics are important foundations for an integrative understanding of lymphatic biology and development. The physical environment includes both biochemical (i.e., ECM composition) and mechanical components, since these are strongly interdependent. Our earlier work with lymphatic function and physiology sought to measure lymph flow and to mathematically describe the dependence of the mechanical environment on lymphatic function.[13] This led us to extend our focus on the mechanical environment from functional to biological implications, and to propose that interstitial stress not only regulates the function, but may also initiate or regulate the development of lymphatic capillaries.

MOUSE TAIL SKIN MODEL FOR LYMPHATIC FUNCTION AND BIOLOGY

We have characterized normal lymphatic function in mouse tail skin and used it in various capacities for multi-level studies of lymphatic function, physiology, and biology. The mouse tail skin exhibits several features that make it conducive to quantitative studies of interstitial–lymphatic transport phenomena. First, it has a very regular, hexagonal network of superficial lymphatic capillaries that are readily visible under the microscope following uptake of fluorescent molecules (FIG. 3a). This technique is called fluorescence microlymphangiography; it involves injecting the solution intradermally into the tail tip and as the fluid tracer gets taken up into the lymphatics, it reveals the functional vessels through which it flows. Also, an injection of fluid trac-

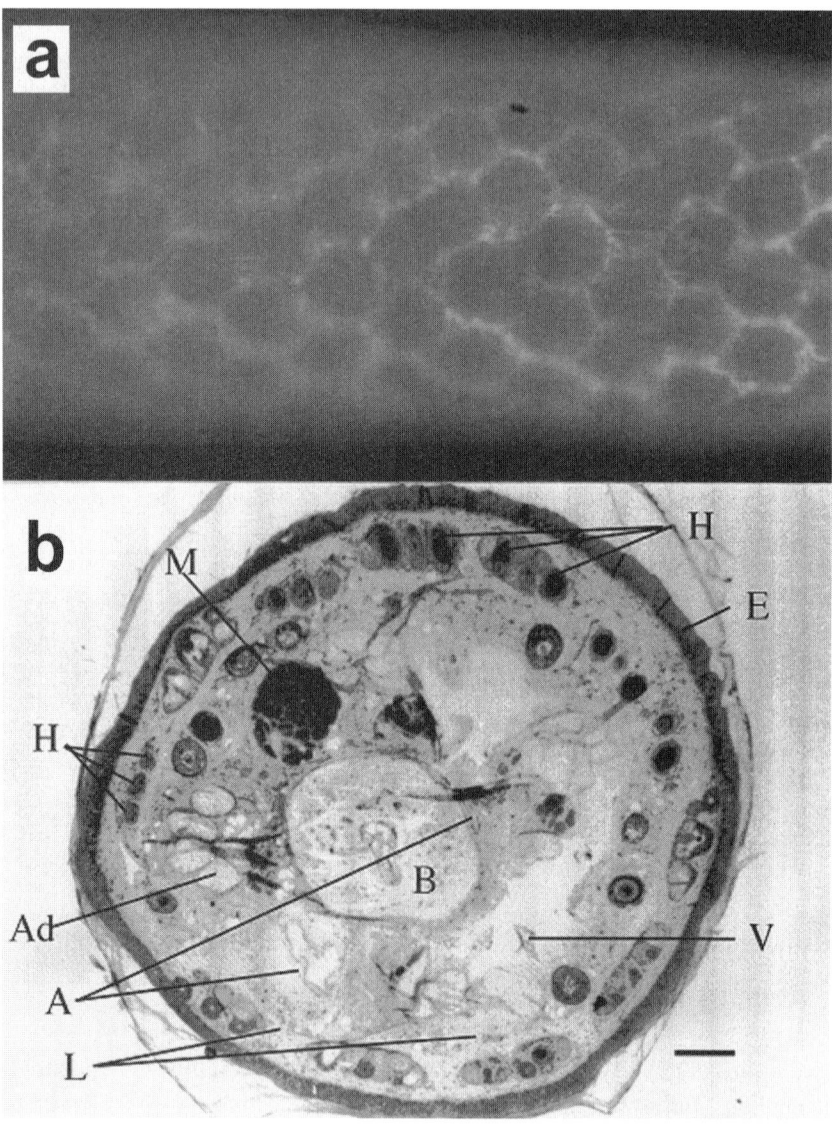

FIGURE 3. (a) Microlymphangiography in the mouse tail, using FITC-Dx 2M (0.0125 g/ml) infused intradermally into the mouse tail, reveals the strikingly regular hexagonal network of lymphatic capillaries that lies just beneath the epidermis. Mesh diameter is approx. 400 mm. (b) Cross-section of a BALB/C mouse tail, stained with Toluidine blue. B, bone; V, venule; L, lymphatic capillary; A, arteriole; Ad, adipose tissue; H, hair follicle; M, skeletal muscle bundle; E, epidermis. Scale bar, 100 mm.

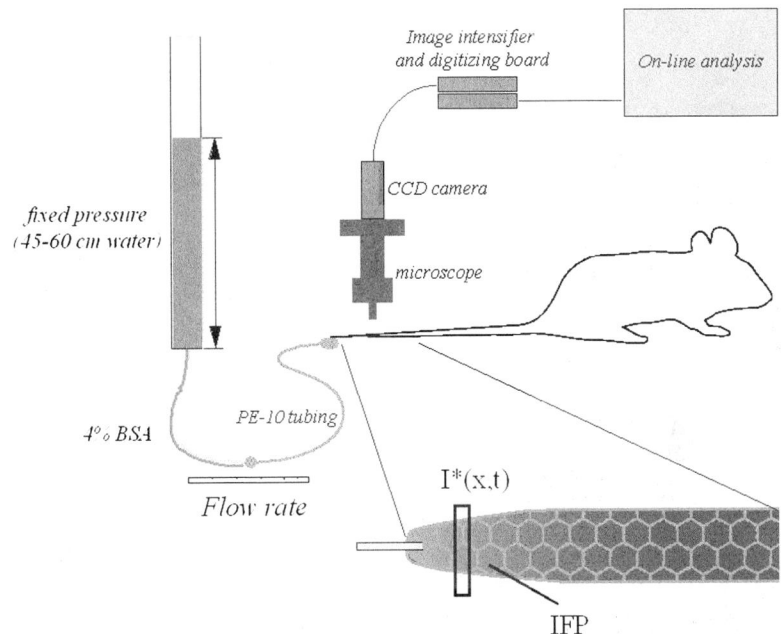

FIGURE 4. Experimental setup for quantifying tissue fluid balance parameters. At time $t = 0$, the infusion begins at fixed pressure into the tip of the tail. Infusion flow rate is monitored while a solution of FITC-dextran is deposited interstitially. Fluid flow resulting from this infusion can be measured either by observing the convecting fluorescent solution or by measuring IFP gradients with micropipettes. (From Swartz et al.[13] Reprinted by permission.)

er at the tail tip allows transport only in the proximal direction (i.e., toward the animal), first within the tissue interstitium and then within the lymphatic network. This is a minimally invasive procedure that requires anesthesia only to immobilize the animal for observation, and the animal can be observed as such repeatedly for several months. Furthermore, the tail is essentially an annulus of dermal and subcutaneous tissue (i.e., skin) surrounding a core of bone and tendons (FIG. 3b); as such, interstitial transport of the intradermally injected material occurs in a fairly homogeneous tissue and mostly in one direction.

This model was initially used to characterize and quantify flow velocity within the lymphatic network following an interstitial infusion of fluorescently labeled dextran at constant pressure (FIG. 4).[28,29] Because of its ideal characteristics for transport studies, we then continued to elaborate the model for quantitative measures of lymphatic function by evaluating mechanical events at the interstitial–lymphatic interface. A theoretical model was developed to describe the balance between lymphatic uptake, fluid flow, and fluid

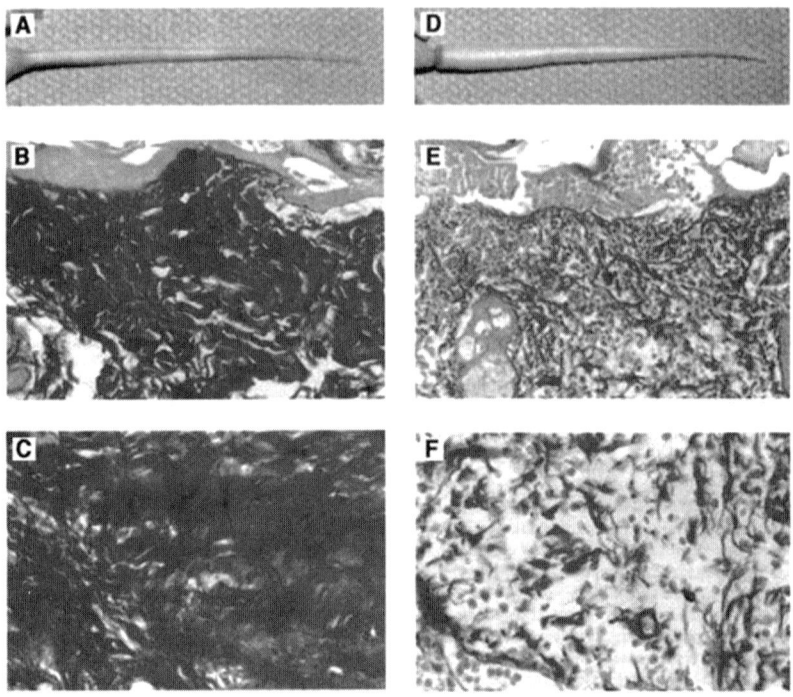

FIGURE 5. Ligation of the deep and superficial lymphatics in the mouse tail skin consistently and reproducibly yielded lymphedema in all subjects. Normal mouse tail (**a**) and edematous mouse tail (**d**) 7 days after ligation. Normal (**b**) and edematous (**e**) tail skin sections stained with a trichrome stain (×40) showing the severe disruption of the ECM. Close-up (×60) sections of normal (**c**) and edematous (**f**) skin. (From Swartz et al.[13] Reprinted by permission.)

pressure in the tail infusion system.[13] By measuring transient and spatial distributions of interstitial fluid pressure (IFP) with micropipettes, we validated the model and estimated three key bulk parameters of tissue fluid balance: lymphatic conductance, hydraulic conductivity, and ECM elasticity. This unique model provided the first *in situ* measurements of these parameters. We also modified this model to study the changes in these fluid transport properties during altered conditions such as chronic lymphedema (FIG. 5).[13]

This same model was then used to integrate functional lymphatic mechanics with developmental lymphatic biology in order to gain insight into their relationship. In transgenic mice with dermal-specific upregulation of VEGF-C, morphological and functional differences in the tail skin lymphatics were compared with those in wild-type animals.[23] The transgenic mice had lymphatic vessels that were hyperplastic, although the capillary spacing and architecture was normal; furthermore, the ability of these hyperplastic and hypertrophic vessels to drain interstitial fluid was significantly decreased (unpublished data). These observations highlighted the possibility of different regulatory mechanisms between (1) lymphatic architecture, organization, and vessel number or density and (2) LEC proliferation. It is also interesting to note that the only differences between these mice were in their expression of VEGF-C, not in the ECM, mechanical environment, or any other factor that would affect tissue fluid balance or lymphatic function. These results are in line with the hypothesis that lymphangiogenesis is a process of interstitial fluid flow management; if lymphatic development and architecture are physiologically regulated by the local mechanical environment then this regulator would lie upstream of molecular regulation such as VEGF-C and other lymphatic growth factors may be produced in response to the mechanical environment.

CDE MODEL FOR OBSERVATIONS OF LYMPHATIC DEVELOPMENT

As described in a recent publication,[30] we developed a model in the mouse tail skin dubbed "collagen dermal equivalent," or CDE, to investigate the hypothesis that lymphangiogenesis is instigated by mechanical stress resulting from interstitial fluid flow. This model uniquely enables us to (1) observe the process of lymphangiogenesis over time, from both functional as well as molecular perspectives; (2) alter the mechanical environment so that the role of interstitial flow can be evaluated; and (3) alter the biochemical environment directly (e.g., with growth factors or tumor cells) to compare their effects to those of mechanical stress alone. Briefly, a mid-tail circumferential portion (2 mm wide) of mouse tail skin is removed and replaced by a collagen gel "window" (FIG. 6a-b). This excised skin contains the lymphatic capillary network, so lymphatic continuity between the distal and proximal ends is disrupted. At any time, we can observe lymph flow—either as interstitial flow or through discrete channels—within the gel using fluorescence microlymphangiography (FIG. 6c). The tracer is lysine-fixable FITC-dextran, which crosslinks to the fluid channel or vessel wall upon fixation so that the fluid channels can easily be identified in thin sections (FIG. 6d), allowing us to spatially correlate functional and molecular information and answer such questions as: Are fluid channels always lined with endothelial cells? If not when does their co-localization occur? Is growth factor production associated with the fluid channels, and if so, in what stages? The model is sustainable for indefinite periods of time; in fact, after a few weeks the collagen becomes integrated with the existing skin (FIG. 6e) and the tail appears normal.

Part of our hypothesis rests on the assumption that fluid channeling is a more efficient way to transport fluid than nonchannel (interstitial) flow. This

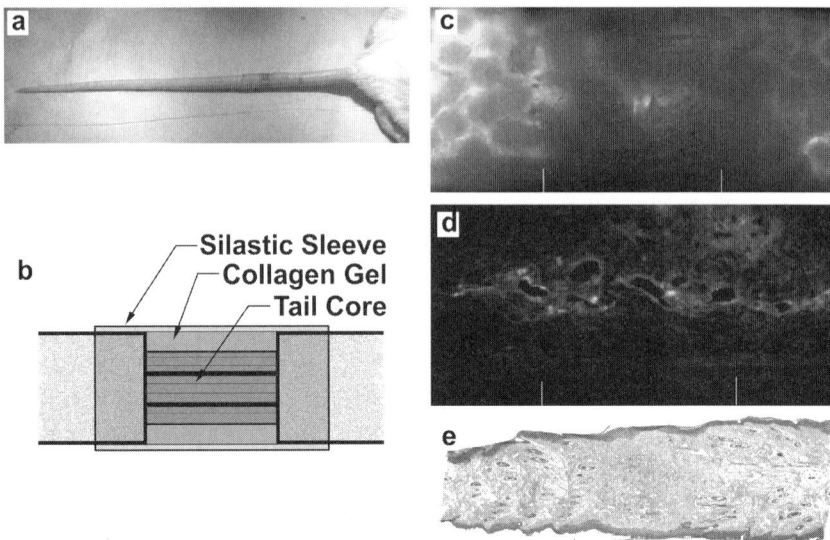

FIGURE 6. (a) Mouse shortly after implantation and (b) schematic of the collagen dermal equivalent (CDE) model. A 2-mm circumferential gap of skin is removed, leaving a core of bone, tendons, blood vessels, and nerves encased in a fascial layer. A silicone sleeve is fitted over the gap and a 0.2% type I collagen solution is injected and allowed to gel. (c) Fluid channels through the gel that are continuous with lymphatic capillaries are made visible by fluorescence microlymphangiography. (d) Upon fixation, the FITC-dextran cross-links to the tissue, permitting identification of those functional channels or lymph vessels in thin sections. Shown here is a section from the same mouse whose functional network is shown above in FIGURE 5c. (e) Standard hematoxylin and eosin staining of a thin section (again, of the same mouse tail) shows that within 25 days, the collagen gel appears to be well integrated with the existing tissue. The only way to identify the gel in sections is by the absence of hair follicles and epidermal glands. A normal epidermal layer, collagen architecture, and interstitial cell population can be seen within the gel, attesting to its biocompatability

warranted a theoretical comparison between fluid flow through the collagen gel and that through fluid channels or lymphatic vessels. For simplicity, we decided to consider the initial and final cases: (1) that of interstitial flow through a homogeneous, acellular collagen gel and (2) that through a fully organized lymphatic network.

First, interstitial flow rates through the collagen gel must be the same as those through the lymphatic network for fluid continuity. Since the gel presumably poses a higher resistance to fluid flow than the vessels, there must be a higher driving force—fluid pressure—to maintain this flow rate. For low Reynolds number flow, the relationship between the pressure drop $\Delta P/L$ and flow Q_{tube} in a straight tube is given by Poiseuille's law:

$$Q_{tube} = \frac{\pi R^4}{8\mu} \frac{\Delta P}{L}$$

where R is the tube radius. Thus, the pressure drop within the lymphatic capillary network in the tail is

$$\left(\frac{\Delta P}{L}\right)_{lymph} = \frac{Q_{tot}}{n}\left(\frac{8\mu}{\pi R^4}\right),$$

where n = average number of vessels per cross-sectional area. In reality, the pressure drop will be higher due to the tortuosity and multiple junctions of the lymphatic network.

The flow through the collagen gel must be the same as that through the lymphatic network to maintain fluid continuity, and the pressure required to drive this flow depends on K, the hydraulic conductivity of the gel (or inverse resistance to flow). According to Darcy's Law, $Q/A = -K\nabla P$ through a homogeneous porous medium, where A is the cross-sectional area to flow (i.e. the gel), and ∇P is the pressure gradient across the gel, $(\nabla P/L)_{gel}$. Therefore the pressure drop required to flow Q_{tot} through the gel is $(\nabla P/L)_{gel} = Q_{tot}/AK$. We used an *in vitro* interstitial flow chamber to estimate K in a 0.2% type I rat tail tendon collagen gel at 37°C and found it to be $2.1 \pm 0.3 \times 10^{-5}$ cm^2/s/mm Hg; the dimensions of the gel are A = 1.8 mm^2 and L = 2 mm. The tail skin contains an average of n = 14 vessels of average diameter 68 mm per any given cross-section.[a] With these estimates, the ratio

$$\frac{(\Delta P/L)_{gel}}{(\Delta P/L)_{lymph}} = 220.$$

This gives us a sense for the difference in the forces required for fluid to flow at a rate Q_{tot} through the gel compared to through the lymphatic capillary network. If the pressure driving lymph flow remains constant, then by replacing the skin with the circumferential CDE, the flow rate of lymph will be reduced by a factor of 200. We previously measured lymph flow velocity in the tail lymphatic network at 3.5 mm/s,[28,29] leading to a net lymph flow rate of roughly 0.01 mL/min. Therefore we can expect the flow rate through the gel initially to be around 5×10^{-5} mL/min (again, assuming the pressure drop driving this flow has not changed). As cells migrate into the gel, remodel the ECM, and form fluid channels, this net flow rate will increase from 5×10^{-5} mL/min (pure collagen gel) to 0.01 mL/min (normal capillary network) as normal functioning is restored.

---

[a]The geometry and dimensions of the tail skin lymphatic network were verified by CT images taken recently by C. Dawson, R. Molthen, and S. Haworth, Biomedical Engineering Department, Marquette University, Milwaukee, WI (data not shown).

This tool provides much insight into the process of lymphangiogenesis. In normal skin, fluorescent channels resulting from microlymphangiography define functional lymphatics. In the CDE, where lymphatics may be absent, developing, or fully formed, the fluid tracer indicates the patterns of fluid flow from distal lymphatics (i.e., this may or may not be within a lymphatic capillary). After sectioning and immunostaining, we can distinguish "fluid channels" from "lymphatic vessels" within the gel only by co-localization of LYVE-1 and Flt-4 positive cells with the green fluorescence. This ability to identify functional lymphatic vessels and differentiate LECs from blood endothelial cells (BECs) through combined functional and molecular techniques is a key advantage, particularly since some putative lymphatic markers such as Flt-4 and LYVE-1 are promiscuous under certain conditions.[31–33]

In a recently submitted manuscript, we describe how this model is used to provide the first evidence that interstitial flow plays a regulatory role in lymphatic development that is upstream from molecular regulators such as VEGF-C. That evidence included observations that fluid channeling precedes LEC migration and lymphatic capillary organization, and that LEC migration and organization occurs only in the direction of flow.[30] Just as angiogenesis is a tissue response to hypoxia, our recent work suggests that lymphangiogenesis is a tissue response to interstitial fluid flow. This demonstrates departure from the parallels that have been tacitly assumed between blood and lymph angiogenesis. Indeed, the organizing principles of blood and lymph angiogenesis are different and are likely to be related to their specific physiological functions despite many similarities at the cellular level.

ACKNOWLEDGMENTS

We are grateful to the Illinois Division of the American Cancer Society and the Department of Defense Breast Cancer Research Program for funding much of the work that is reviewed here.

REFERENCES

1. AUKLAND, K. & R.K. REED. 1993. Interstitial-lymphatic mechanisms in the control of extracellular fluid volume. Physiol. Rev. **73:** 1–78.
2. SCHMID-SCHÖNBEIN, G.W. 1990. Microlymphatics and lymph flow. Physiol. Rev. **70:** 987–1028.
3. LEAK, L.V. & J.F. BURKE. 1966. Fine structure of the lymphatic capillary and the adjoining connective tissue area. Am. J. Anat. **118:** 785–810.
4. LEAK, L.V. & J.F. BURKE. 1968. Ultrastructural studies on the lymphatic anchoring filaments. J. Cell Biol. **36:** 129–149.

5. IKOMI, F. & G.W. SCHMID-SCHÖNBEIN. 1996. Lymph pump mechanics in the rabbit hind leg. Am. J. Physiol. **271:** H173–83.
6. SCHMID-SCHÖNBEIN, G.W. 1990. Mechanisms causing initial lymphatics to expand and compress to promote lymph flow. Arch. Histol. Cytol. **53** Suppl.: 107–114.
7. GRANGER, H.J. 1981. Physicochemical properties of the extracellular matrix. *In* Tissue Fluid Pressure and Composition. A.R. Hargens, Ed. :43–61. Williams & Wilkins. Baltimore, MD.
8. GRODZINSKY, A.J. 1983. Electromechanical and physicochemical properties of connective tissue. Crit. Rev. Biomed. Eng. **9:** 133–199.
9. HARGENS, A.R., Ed. 1981. Tissue Fluid Pressure and Composition. Williams & Wilkins. Baltimore, MD.
10. LEVICK, J. 1987. Flow through interstitium and other fibrous matrices. Quart. Rev. Exp. Physiol. **72:** 409–438.
11. KENEDI, R.M., T. GIBSON & C.H. DALY. 1965. Bio-engineering studies of the human skin. *In* Biomechanics and Related Bio-Engineering Topics. R.M. Kenedi, Ed. :147–158. Pergamon Press. Oxford.
12. MARKENSCOFF, X. & I.V. YANNAS. 1979. On the stress-strain relation for skin. J. Biomech. **12:** 127–129.
13. SWARTZ, M.A. *et al.* 1999. Mechanics of interstitial-lymphatic fluid transport: theoretical foundation and experimental validation. J. Biomech. **32:** 1297–1301.
14. BUDGE, A. 1880. Über lymphherzen bei hühnerembryonen. Arch. Anat. Entwickl.-Gesch. :350–359.
15. SALA, L. 1900. Sullo sviluppo dei cuori linfatici e dei dotti torici nell' embryone di pollo. Ric. Lab. Anat. Norm. Univ. Roma. **7:** 899–1000.
16. OHTA, Y., V. SHRIDHAR, R.K. BRIGHT, *et al.* 1999. VEGF and VEGF type C play an important role in angiogenesis and lymphangiogenesis in human malignant mesothelioma tumours. Br. J. Cancer. **81:** 54–61.
17. KARPANEN, T. 2001. Vascular endothelial growth factor C promotes tumor lymphangiogenesis and intralymphatic tumor growth. Cancer Res. **61:** 1786–1790.
18. MANDRIOTA, S.J. 2001. Vascular endothelial growth factor-C-mediated lymphangiogenesis promotes tumour metastasis. EMBO **20:** 672–682.
19. SKOBE, M., *et al.* 2001. Induction of tumor lymphangiogenesis by VEGF-C promotes breast cancer metastasis. Nature Med. **7:** 192–198.
20. PEPPER, M.S. 2001. Lymphangiogenesis and tumor metastasis: myth or reality? Clin. Cancer Res. **7:** 462–468.
21. JOUKOV, V. *et al.* 1996. A novel vascular endothelial growth factor, VEGF-C, is a ligand for the Flt4 (VEGFR-3) and KDR (VEGFR-2) receptor tyrosine kinases. EMBO J. **15:** 1751.
22. LEE, J., *et al.* 1996. Vascular endothelial growth factor-related protein: a ligand and specific activator of the tyrosine kinase receptor Flt-4. Proc. Natl. Acad. Sci. USA. **93:** 1988–1992.
23. JELTSCH, M. *et al.* 1997. Hyperplasia of lymphatic vessels in VEGF-C transgenic mice. Science **276:** 1423–1425.
24. OH, S.J., *et al.* 1997. VEGF and VEGF-C: specific induction of angiogenesis and lymphangiogenesis in the differentiated avian chorioallantoic membrane. Dev. Biol. **188:** 96–109.

25. JOUKOV, V. et al. 1997. Proteolytic processing regulates receptor specificity and activity of VEGF-C. EMBO J. **16:** 3898–3911.
26. ACHEN, M.G., et al. 1998. Vascular endothelial growth factor D (VEGF-D) is a ligand for the tyrosine kinases VEGF receptor 2 (Flk1) and VEGF receptor 3 (Flt4). Proc. Natl. Acad. Sci. USA. **95:** 548–53.
27. STACKER, S.A., et al. 2001. VEGF-D promotes the metastatic spread of tumor cells via the lymphatics. Nature Med. **7:** 186–191.
28. BERK, D.A. et al. 1996. Transport in lymphatic capillaries: II. Microscopic velocity measurement with fluorescence recovery after photobleaching. Am. J. Physiol. **270:** H330–7.
29. SWARTZ, M.A., D.A. BERK & R.K. JAIN. 1996. Transport in lymphatic capillaries: I. Macroscopic measurment using residence time distribution analysis. Am. J. Physiol. **270:** H324–329.
30. BOARDMAN, K.C. & M.A. SWARTZ. Interstitial flow as a guide for lymphangiogenesis. Submitted for publication.
31. CARREIRA, C.M. et al. 2001. LYVE-1 is not restricted to lymph vessels: expression in normal liver blood sinusoids and down-regulation in human liver cancer and cirrhosis. Cancer Res. **61:** 8079–8084.
32. VALTOLA, R. et al. 1999. VEGFR-3 and its ligand VEGF-C are associated with angiogenesis in breast cancer. Am. J. Pathol. **154:** 1381–1390.
33. PARTANEN, T.A. et al. 2000. VEGF-C and VEGF-D expression in neuroendocrine cells and their receptor, VEGFR-3, in fenestrated blood vessels in human tissues. FASEB J. **14:** 2087–2096.

Proteomic Technologies to Study Diseases of the Lymphatic Vascular System

LEE V. LEAK,[a,b,c] EMANUEL F. PETRICOIN, III,[b] MICHAEL JONES,[c] CLOUD P. PAWELETZ,[a,b] ALI M. ARDEKANI,[b] VINCENT A. FUSARO,[b] SALLY ROSS,[b] AND LANCE A. LIOTTA[a]

[a]*Laboratory of Pathology, National Cancer Institute, National Institutes of Health, Bethesda, Maryland 20892, USA*

[b]*Clinical Proteomics Program of Therapeutic Proteins, CBER, Food and Drug Administration, Bethesda, Maryland 20892, USA*

[c]*Department of Anatomy, College of Medicine, Howard University, Washington, DC 20059, and the Laboratory of Animal Surgery and Medicine, National Heart Lung and Blood Institute, National Institutes of Health, Bethesda, Maryland 20892, USA*

ABSTRACT: Now that the human genome has been mapped, a new challenge has emerged: deciphering the various products of individual genes. Consequently, new proteomic technologies are being developed to monitor and identify protein function and interactions responsible for the total activities of the cell. The application of these new proteomic technologies to study cellular activities, will lead to a faster sample throughput and increased sensitivity for the detection of individual proteins, thus providing major opportunities for the discovery of new biomarkers for the early detection of protein alterations associated with the progression of the disease state.

KEYWORDS: proteomics; cancer; lymph; lymphatic endothelium; lymphedema; laser capture microdissection; protein microarrays; SELDI-TOF-mass spectroscopy

INTRODUCTION

The lymphatic vascular system plays a vital role in the overall function of the cardiovascular system through the removal of permeated plasma proteins from the interstitium and their return to the systemic blood circulation. The maintenance of physiologic fluid homeostasis in higher vertebrates depends on this constant return. In various diseases of the lymphatic vascular system, the normal return of permeated plasma proteins and lymphocytes to the systemic circulation is hampered or completely blocked, leading to severe lymphedema.[1,2] This scenario occurs in a variety of lymphatic disorders, such as lymphangiodysplasia syndromes, Milroy's lymphedema, chylous reflux with peripheral edema, lymphatic dysfunctions associated with cancer, and primary and secondary lymphedema subsequent to surgical interventions, yet the puzzles posed by many of the critical diseases that involve the lymphatic system remain unsolved.[2–4] Recent studies using the methods of molecular genetics have shown that various genetic mutations are associated with various hereditary diseases of the lymphatic vascular system that produce different types of lymphedema.[4,5] In addition, key scientific advances have also come from the identification and characterization of the growth factor VEGF-C, which is a potent mitogen that acts on the lymphatic endothelium and results in the formation of new lymphatic vessels, in addition to affecting the permeation of the blood vascular system.[6–11]

The completion of the Human Genome Project is providing bountiful information regarding the sequences of genes. Studies estimate that the human genome encodes between 30,000 and 40,000 structural genes.[12,13] However, due to post-transcriptional and post-translational modifications, such as phosphorylation, glycosylation, and proteolysis of proteins, the number of proteins expressed by cells is believed to far exceed these estimates.[12,13] A major challenge in biology and medicine is to focus on the immense task of identifying the structure, function, and interactions of proteins produced by individual genes and the genetic mutations that lead to specific disease processes. Although the genome contains the information archive that directs the production of all cellular products (proteins and lipoproteins), proteins carry out the work of the cell and ultimately orchestrate its biological activities and fate. Therefore, detection and analysis of protein profiles, to determine the identity and relative quantity of proteins and their isoforms in normal and pathological states, represent some major challenges of the post-genomic era. Thus, a primary focus of lymphatic research must now be directed toward the tremendous task of identifying and characterizing the array of proteins produced by individual genes that carry out the work and activities of the lymphatic endothelium.

The global expression of proteins by the cell is known as the proteome, and their analysis constitutes the field of proteomics. It consists of the identification and quantitation of proteins and the investigation of protein structure, localization, modification, interaction, and function.[14–17] Bringing proteomic analysis to bear on diseases of the lymphatic vascular system will provide new ways to examine both physiological and pathological processes at the molecular level, providing new insights for the various processes leading to dysfunction of the lymphatic system. This presentation will discuss new proteomic technologies for protein analysis in cancer, and the application of proteomics in studies of protein profiles in lymph and in the lymphatic endothelium.

PROTEOMIC TECHNOLOGIES

Laser-Capture Microdissection

A major challenge in proteomics is to define the state of proteins in the cell, within their natural context. Since no cell exists or functions in isolation, the microenvironment within a given tissue or organ system represents an important factor in analyzing and characterizing the activities of a given cell type. Cell–cell contact, autocrine function, paracrine growth factors, blood vascular circulation and lymphatic drainage, and external forces have all been shown to affect cellular activity and function. While *in vitro* models, such as co-cultures of different cell types[18] and *ex vivo* tissue culture in bioreactors,[19] have been established to represent cells *in vivo*, the results from these studies indicated that primary cells lose their differentiated state in a short period of time when grown in culture and that their metabolism is significantly changed in continuous cultures.[20,21] These studies indicate that primary cells should be used to analyze protein changes in the disease state. To this end, tools are currently available that allow the direct isolation of a specific cell type from intact tissues, without the necessity of cell disruption. One such technology used in our laboratory is laser-capture microdissection (LCM). This approach represents one of the advanced technologies that allow the isolation of selected cells types from stained tissues.[22,23] The specimen is placed on a standard microscopic slide and appropriately fixed and stained. The slide is placed onto the LCM stage and a region of interest is delineated. LCM takes advantage of an infrared laser that melts a thermosensitive ethylene vinyl acetate polymer film that is placed over the designated single cell of interest. After the capture of one to thousands of the specific cell types, they are lifted from the remaining tissue section. The laser capture does not affect the enzymatic activity, three-dimensional conservation, or post-translational modification of proteins or the nucleic acids. Therefore, the laser-captured cells can be analyzed for their DNA, RNA, and protein content (FIG. 1). This simple system provides a method for researchers to obtain specific cell types from complex heterogeneous tissue by isolating single cells under direct microscopic visualization. The impact of this technology has been impressive, with its ability to provide pure samples of specific cells such as normal and cancerous cells. With LCM technology, protein expression profiles have been analyzed in normal and diseased epithelium, as well as in the surrounding stroma and in premalignant lesions.[24–26] To determine the protein profiles of

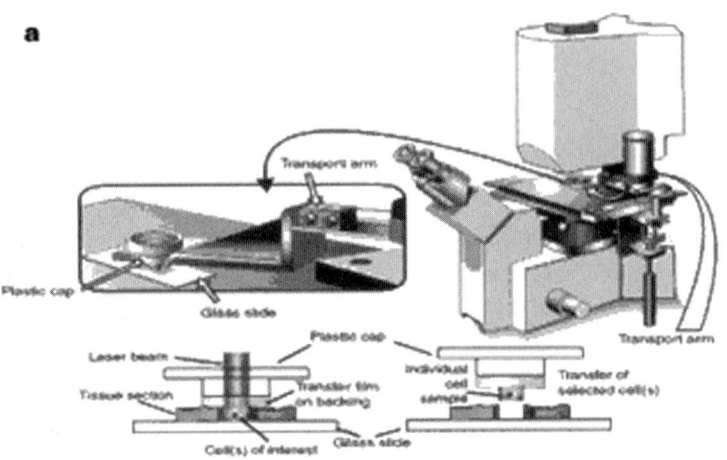

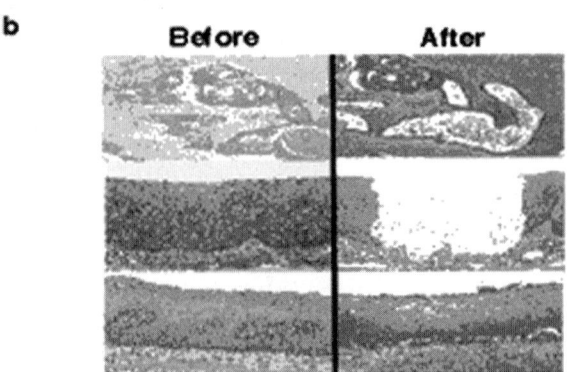

FIGURE 1. (**a**) Schematic overview of laser-capture microdissection. The laser-capture microdissection microscope consists of an inverted pathologic microscope, an infrared laser, and a cap housing an infrared labile film. Upon firing of the laser, the transfer film melts into the voids of the tissue and rapidly cools down. Cells are isolated when the cap is lifted. (**b**) Esophageal section (hemotoxylin and eosin–stained) before and after cell isolation by laser-capture microdissection. (From Simone *et al.*[59] Reproduced by permission.)

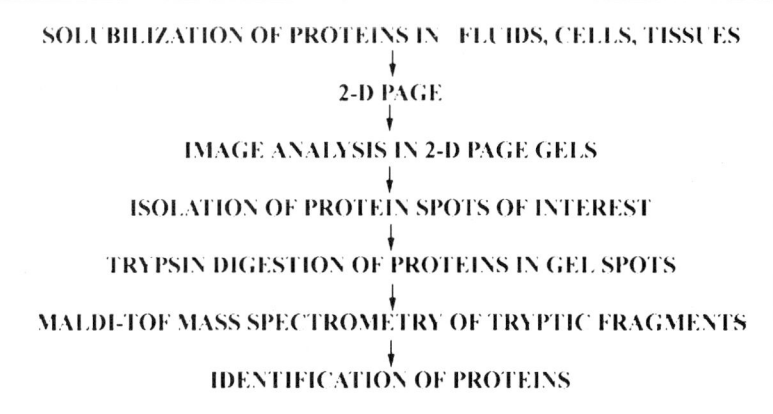

FIGURE 2. Diagram showing an outline of a general approach for the separation, visualization, and identification of complex protein mixtures. (Adapted from Chambers et al.[60])

specific cell types obtained by LCM, we have employed the methods of high-resolution two-dimensional polyacrylamide electrophoresis (2-D PAGE), as described below.

Traditionally, the dominant approach in protein separation and analysis has combined high resolution 2-D PAGE with mass spectrometry (MS) for the identification of both amino acid sequence and post-translational modification of the fractionated proteins from complex mixtures of cells and tissues.[27,28] 2-D PAGE for the separation of protein mixtures is a powerful tool for surveying biological complexity at the molecular level.[29] This approach has been used extensively to study changes in protein expression in cells and tissues.[30] Through the use of this technology, integrated with other methods of protein analysis, such as mass spectrometry, thousands of proteins can be separated and identified in an automated fashion, through sequential analysis of peptide mixtures that have been generated by digestion of individual gel spots separated in 2-D PAGE gels.[31] The overall strategy for the separation and visualization of complex protein mixtures for proteomic analysis is shown in FIGURE 2. With the methods of 2-D PAGE, proteins are separated by charge (isoelectric focusing [IEF]) in the first dimension and by mass (SDS PAGE) in the second dimension. Although SDS-PAGE and one-dimensional IEF resolve approximately 100 proteins in a heterogeneous sample, through the combination of IEF and SDS-PAGE in a 2-D separation, it is possible to achieve a theoretical resolution of 10,000 or more individual protein spots in a single 2-D PAGE gel.[32] Once proteins are separated and detected in 2-D PAGE gels, a desired outcome of gel imaging in normal vs. diseased samples is the identification of differences that can be related to metabolic changes or alterations produced by the disease. Proteins have been characterized using size, isoelectric point, the presence of modifications, solubility, and start and finish (N- and C-terminus). Traditionally the start of a protein was determined by Edman sequencing. While this procedure is slow and expensive, sufficient information can be obtained to identify the protein, based on a short sequence (4–8 amino acids), provided that the protein is not N-terminally blocked. Through improvements in blotting technologies,[33] it has become possible to sequence proteins from 2-D PAGE gel blots. Proteins are also being identified from 2-D PAGE gel spots on the basis of their amino acid composition and matching to the composition of proteins in databases.[34] To the present, mass spectrometry (MS) is overwhelmingly employed as the method of choice for protein identification from 2-D PAGE gels. Technological progress has led to the development of "soft" methods of ionization, such as matrix-assisted laser desorption/ionization (MALDI) and electrospray ionization (ESI). These advances have made it possible to obtain the primary sequence information of peptides and proteins.[35] [See Refs. 15 and 36, reviews that discuss specific techniques and instruments of the tools employed in the identification of separated protein species]. In addition, the development of a database of DNA (protein sequence) information has made it possible to use MS methodology for protein identification in large-scale proteomic projects. In this approach, theoretical tryptic digests derived from sequence databases are compared with experimental digests. Therefore, by combining 2-D PAGE and MS, several thousand protein species can be separated, detected, and quantified. This can be followed by the detection and identification of hundreds of proteins in a highly automated fashion through sequential analysis of the peptide mixtures that can be generated by digestion of individual protein gel spots.[31] The protein of interest is enzymatically or chemically cleaved and aliquots of the obtained mixture are analyzed by mass spectrometric techniques. Proteomic Work Systems, with software packages, are currently available commercially. These provide a high throughput for 2-D PAGE gel analysis and subsequent gel spot cut-out for enzymatic protein digestion and MALDI-TOF mass spectrometric analysis. Peptide masses are then used to determine the peptide fingerprint, and identification of the protein is accomplished after comparison of the resolved spectra with theoretical protein and peptide masses within databases.

By combining the traditional methodology for proteomic analysis with one of the new technologies, such as LCM, it is possible to routinely isolate specific cell types for proteomic analysis in normal tissues and within various pathological conditions, including the progression of cancer. LCM technology has also been used to obtain information on the protein changes in normal and cancerous epithelial cells from a single patient's tissues in early stages of tumorigenesis. In FIGURE 3, microdissected cell–patient-matched normal ep-

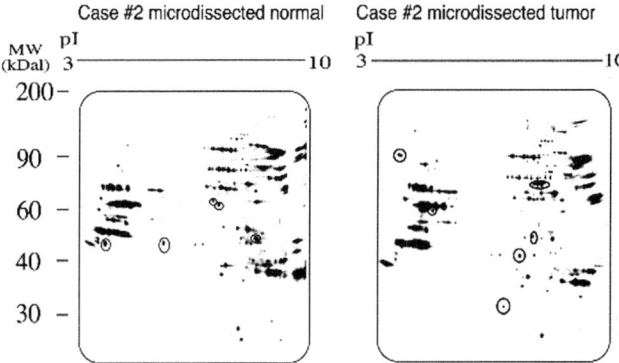

FIGURE 3. Two-dimensional polyacrylamide gel electrophoresis (2-D PAGE) comparison of normal epithelial and pure tumor cells from the same patient's prostatectomies: 35,000–50,000 cells were obtained via LCM-transfer, directly lysed in IEF buffer, and run on a pH 4–7 Pharmacia IPG IEF gel strip for 100kVh. The second-dimensional run was performed on 9% isocratic SDS-PAGE gels and the gels were stained with silver. The normal and tumor gels represent a protein fingerprint for each gel. (Adapted from Ornstein et al.[24])

ithelium, tumor cell populations were applied to 2-D PAGE and analyzed. Using software to analyze the relative intensity of protein spots in silver-stained 2-D PAGE gels, it was determined that the overall protein patterns differed among normal and tumor cells.[37] By excising novel protein spots from 2-D PAGE gels and sequencing them by mass spectrometry, it has been possible to identify proteins that abound differentially in normal vs. premalignant vs. invasive breast carcinoma, prostate carcinomas, ovarian carcinomas, and esophageal carcinomas.[38] This represents the first application of proteomics directly to the tissue microenvironment to microdissect a homogeneous population of specific cell types. In these studies, the proteomic pattern was found to be 80% different in the microdissected tissue cells when compared to cultured cells from the same patient's tissue.[38] The detection of such specific protein changes, coupled with their localization in specific cells types, is a powerful method for the discovery of biomarkers associated with early alterations in proteins in the progression of cancer.

The methods of proteomics are also being applied to studies of lymph and the lymphatic endothelium. The differentiated state of the endothelium is represented by a tubular vessel, both in the blood circulation and in lymphatics. In vivo, the formation of new vessels, blood (angiogenesis) and lymphatic (lymphangiogenesis), occurs from existing vessels.[39–41] Studies have also shown the sprouting from existing endothelial lining of segments of thoracic ducts embedded in plasma clots in vitro.[42] Studies in our laboratory and those of others have described the formation of tubular channels in monolayer cultures of lymphatic endothelial cells, a cell line derived from tumors, and a transformed endothelial cell line.[43–46] Although it is known that lymphatic vessels develop and regenerate later than blood vessels during embryonic development, very little is known about protein expression changes when the lymphatic endothelium undergoes phenotypic changes to accommodate the formation of new lymphatic vessels. Techniques are now available that will permit the isolation of homogenous populations of lymphatic endothelial cells (LECs), to place them in a defined medium for cell culture.[47,48] The various methods of cell culture have proven to be of great advantage, for their ability to allow the process to be monitored, to allow continuous manipulation of the experimental conditions, and to permit knowledge of the precise parameter of the culture system. In applying high-resolution 2-D PAGE to determine the feasibility of analyzing protein profiles of the lymphatic endothelium during the process of lymphangiogenesis, we have been able to detect major differences in protein profiles between the quiescent LEC and LEC undergoing tube formation *in vitro*. This has been accomplished by comparing protein profiles in silver-stained 2-D PAGE gels of LEC monolayer cultures and LECs in lymphangiogenesis. The proteomic pattern comparison shows a marked difference in specific protein changes in the sprouting vessels with absence of these phenotypic changes in LEC monolayer cultures. The exciting finding to date is that there are a number of unique changes in protein expression profiles that appear to occur in association with the morphological correlates. By subjecting the novel protein spots observed in 2-D PAGE gels to mass spectrometric analysis, we can better understand those protein components which serve as key functional players to effect the biological activities of the lymphatic endothelium during the process of lymphangiogenesis. Similar proteomic studies are also being carried out with lymph to determine the feasibility of identifying and characterizing proteins that may be unique to lymph when compared to plasma. In addition, this approach may allow the identification of proteins that might be associated with disease, an inflammatory state, a disruption in the flow of lymph, or an accumulation of proteins associated with a carcinogenic event. FIGURE 4 shows 2-D PAGE silver-stained gels of lymph and plasma following cardiopulmonary bypass. While lymph shows many protein spots that are similar to those of plasma, there are also qualitative and quantitative differences. Studies currently under way should provide new insights into the protein alterations that relate to and affect the formation and propulsion of lymph for the maintenance of homeostasis for tissues of the body.

Although high-resolution 2-D PAGE has been the most effective and widely used technology for protein separation, there are still several technical limitations to these methods of protein separation. The inadequate separation of hydrophobic proteins still poses a limitation.[14] Another serious drawback is the inability of 2-D PAGE to detect low-abundance proteins, since there is no

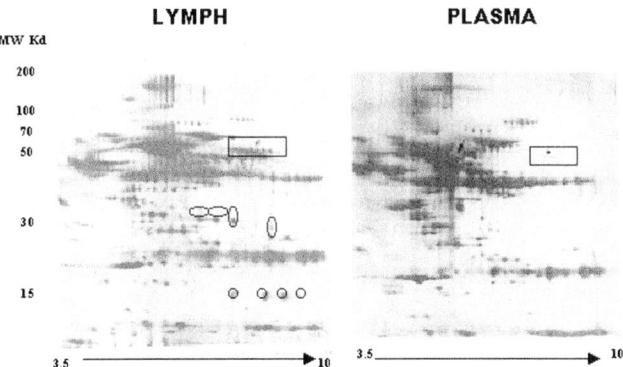

FIGURE 4. Two-dimensional polyacrylamide gel electrophoresis (2-D PAGE) used in comparison of lymph and plasma from the same sheep during cardiopulmonary bypass. Lymph and plasma samples were directly lysed in IEF buffer and run on a 3-10-NL Pharmacia immobilized pH gradient IPG gel strip for 75kVh. The second-dimensional run was performed on 9–18% gradient SDS-PAGE gels and the gels were subsequently stained with silver. The gel for lymph shows major protein spots for fibrinogen (* rectangle), while the gel for plasma contains very little staining for fibrinogen. Proteins found to be differentially expressed in lymph during cardiopulmonary bypass are *circled.*

equivalent in protein biochemistry to PCR technology that would allow for the enrichment of proteins. As described below, the advances made in other methods, along with the development of new technologies, will help to overcome some of these limitations.

PROTEIN MICROARRAY TECHNOLOGY

Increasing numbers of clinical trials have demanded high-throughput proteomic technologies for the rapid processing of micro-quantities of human samples. However, the conventional protein analytical methods, such as 1-D and 2-D PAGE, require the input of large amounts of specimen, in addition to the complex, labor-intensive, and time-consuming processes for sample preparation.[49,50] To overcome these difficulties, researchers have sought to modify traditional technologies or to develop new technologies in order to apply proteomic analysis to profile the functional state of protein networks in cells and tissue components during normal physiological conditions and various disease processes, as required for clinical trials. Protein microarrays represent one of the new technologies for protein analysis. This method is sensitive and precise, and can be designed to profile specific networking processes of the cell's complex proteins. Microarrays can be designed to include a plurality of bait surfaces of labeled ligands that are able to recognize the activated or bound state of a given protein or component within the complex.[51] Protein microarrays are like gene microarrays, except that a protein bait is applied to the surface. The bait can be a small molecule (peptide or drug), multiple antibodies, a recombinant protein, phage display system, or nucleic acids capable of displaying hundreds or thousands of binding events in a single operation.[38,52,53] While hundreds of antibodies can be placed on the solid phase, the binding affinity for each antibody of the dynamic curve may be different, excluding it from the linear portion of the dynamic curve, and the proteins to be identified may not be labeled to the desired or optimum levels for a reliable analysis of the protein lysate. To overcome these limitations in arrays, a different kind of microarray has been developed that uses immobilized cellular protein lysates.[54] The protein lysate may include microdissected normal epithelium, premalignant lesions, and cancer cells, all from the same patient, on the same array. These different protein lysates from the same patient can be treated with specific antibodies. Because protein microarrays can be designed to contain a plurality of bait surfaces or labeled ligands, the protein arrays can be incubated with antibodies designed to profile the functional state of specific known signal pathways, or for defined networks within cells or tissues, to determine whether a specific protein has been activated in its physiological state. Such protein lysates can be produced at a very low cost, with thousands of samples on each array. They are extremely sensitive, to less than one cell equivalent per lysate spot. Although the spot area is 100 times larger than the diameter of a cell, it has been possible to obtain a very good coefficient of variance for inter- and intra-spot reproducibility ranges.[54] The sensitivity of the system is 1,000 molecules, with two standard deviations, above background. Therefore, each individual sample can be arrayed in a miniature dilution curve to facilitate accurate quantification and to expand the dynamic range. This new proteomic technology has provided the opportunity to profile the state of specific known signal pathways within tissue cells, either cultured cells or populations of cells obtained by laser capture microdissection (FIG. 5). One of the signaling pathways studied with protein microarrays was the apoptotic pathway. This is a cascade of phosphorylated and cleaved proteins that ultimately degrades the DNA molecules of the cells. By producing antibodies against different states in this cascade, it was possible to analyze and determine whether the cascade is activated in a microdissected cell population, before and after treatment with a chemotherapeutic agent. Within a few hours after treatment, there was a surge in caspase-3 cleavage, showing the activation of the apoptosis pathway, which indicates that this is a very early event that can be monitored, because of the sensitive nature of protein microarrays. We have also analyzed microdissected normal epithelium, premalignant lesions, invasive carcinoma and the surrounding stroma, all from the same patient and the same lesion. Arraying these samples on the miniature dilution curves provided a very good dynamic range. To de-

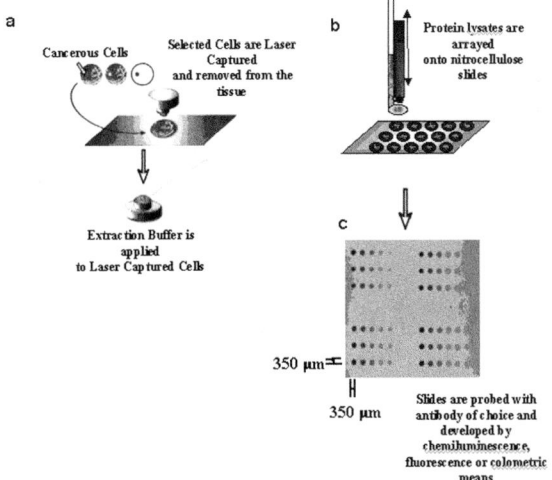

FIGURE 5. Illustration of reverse phase protein array technique (RPPA). (**a**) Defined patient-matched cell populations are microdissected, lysed, and (**b**) arrayed onto nitrocellulose slides with a pin-based microarrayer at distinct positions. (**c**) Each patient set is designed so that it contains a longitudinal cancer progression vertically, and corresponding dilution curves to each disease state horizontally. After arraying, the slides are incubated with an antibody that is detected by chemiluminescent, fluorescent, or colorimetric assays. Intensity of the signal is proportional to the concentration of the target protein. (From Paweletz et al.[54] Reproduced by permission.)

termine whether the apoptotic pathway was activated or suppressed in premalignant lesions of the prostate, protein lysates of prostate cells were applied to microarrays and analyzed using antibodies against components of the apoptotic cascade, as well as the regulators of the apoptotic cascade. These studies showed a big surge in phosphorylation of AKT and a suppression of the apoptotic pathway in the premalignant lesion, even when there is downregulation or a sustained phosphorylation of ERK (FIG. 6). Thus, by combining the proteomic technologies of laser-capture microdissection and protein microarrays, it is possible to use cells progressing to pathology in 10 different patients on one slide to study a specific hypothesis about the evolution of cancer lesions in their microenvironment.[54] In this example, it was shown that premalignancy is associated with the suppression of apoptosis, and that the accumulation of cells might be due to a reduction in cellular death rate, and not to an increase in cell replication rates. Likewise, a suppression of apoptosis facilitates the survival of the tumor cells as they migrate and exit from their site of origin to invade the surrounding tissues. Owing to the suppression of the apoptotic pathway, these cells have escaped the death pathway

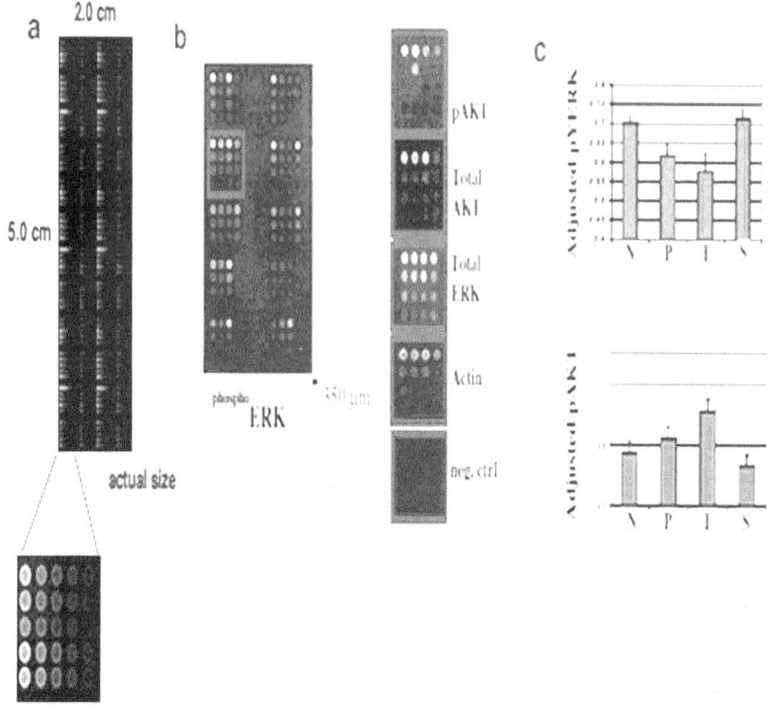

FIGURE 6. Example of reverse phase protein arraying. (**a**) Relative dimensions of RPPA. Eight individual microdissected cases were arrayed in octuplets on one 20 × 50 mm slide yielding 1200 individual array features. (**b-c**) Analysis of cell survival and growth regulation in prostate cancer progression by RPPA. (**b**) A p-Tyr-Erk stained microarray for ten individual cases is shown. Each sector is from an individual patient containing microdissected normal (N), pre-malignant (P), and invasive carcinoma (T) epithelial cells, as well as adjacent stroma cells (S) (vertical column diluted 1-fold, 2-fold, 4-fold, and 8-fold). (**c**) Corresponding RPPA staining of p-SER(473)-Akt, total Akt, total Erk, actin, and negative ctrl of one sector. (Adapted from Paweletz et al.[54])

(apoptosis) and will survive by continuing to replicate in tissues distant from their site of origin. With this proteomic technology, it has been possible to microdissect invading carcinoma cells, the stroma adjacent to the invading carcinoma cells, and the distant stroma, far away from the invading carcinoma cells. Therefore, one can examine specific hypotheses about potential crosstalk between the tumor cells and the host that might regulate the local microinvasion environment. Some examples might include certain proteolytic enzymes (metalloproteineases) that degrade the extracellular matrix: these may be present in cells adjacent to the invading cancer cells or on membranes of the cancer cells. In other cases, cell membrane-associated metalloproteinases may become activated to degrade the extracellular matrix, to facilitate the migration and invasion of the tumor cells. This again demonstrates that protein microarrays can be used to study the microenvironment and develop hypotheses about crosstalk at the proteomic level. With this new proteomic technology, lysates from microdissected tissue cell proteins or cultured cells are immobilized on microarrays. Using unlabeled protein from one biopsy can yield a large number of arrays. Core biopsies from patients enrolled in clinical trials can be immobilized on microarrays to analyze the effects of drugs on an individual patient's lesion. One biopsy can yield hundreds of arrays. This very sensitive method is under current employ at the National Cancer Institute for a number of clinical cancer trials.

PROTEIN CHIP TECHNOLOGY

Both serum and lymph proteomes contain a population of thousands of complex proteins, peptides, low molecular weight proteins, and cleaved peptides that represent a vast, unexplored archive of information. Since all tissues of the body are continuously perfused by serum and lymph proteomes, the protein composition will fluctuate according to the physiological or pathological state of the body and should therefore serve as a barometer reflecting this condition at any given time point. Each cancerous lesion is produced in the context of its unique host/organ microenvironment, that is, the interaction of the cancer cells with the surrounding stroma, endothelial cells, lymphatic vessels, inflammatory cells and host stromal fibroblasts constitutes, in sum, a microenvironment that amplifies the proteomic modifications that eventually appear in serum and lymph. It is not simply a case of looking for proteomic alterations that emanate from the cancer cells directly. The analytic procedures must also take into account the total microenvironment, perfused by lymph and/or blood, that may also be modified by the evolving cancer. In order to discover the diagnostic proteomic pattern in serum, lymph and cell lysates, without knowing the identity of the proteins beforehand, we have used the rapid high-throughput mass spectroscopic method of surface-enhanced laser desorption ionization time-of-flight (SELDI-TOF) MS (Ciphergen Biosystems, Inc., Fremont, CA). This system is capable of generating large amounts of data for analysis in a short time.[55] The SELDI-TOF technology uses an aluminum-based solid support with predefined bait regions that contain different surface-binding chemistries (e.g., hydrophobic, normal-phase, metal-affinity, cationic and anionic) as bait in an area of 1 to 2 mm in diameter.[56] This makes it possible for cell lysates or body fluids in volumes as low as 0.5 µL to be directly applied to these surfaces without the need for prior purification or protein fractionation. After a wash step, the proteins and peptides are selectively retained on the protein chip, based on the bait used and the inherent chemistry of each protein. The chip is placed in a vacuum, and analyzed by MS TOF technology.[55,57] Within the vacuum chamber, the specimen is ionized with a laser. The ionized proteins and peptides are recorded as they strike a detector in a time-dependent fashion, with the small molecules traveling faster and the larger ones taking a longer time to reach the detection plate.[55] A fingerprint of proteins is generated based on the combined, precise molecular weight signatures of the protein complexes or peptides that were bound to the chip and ionized off the specific bait sur-

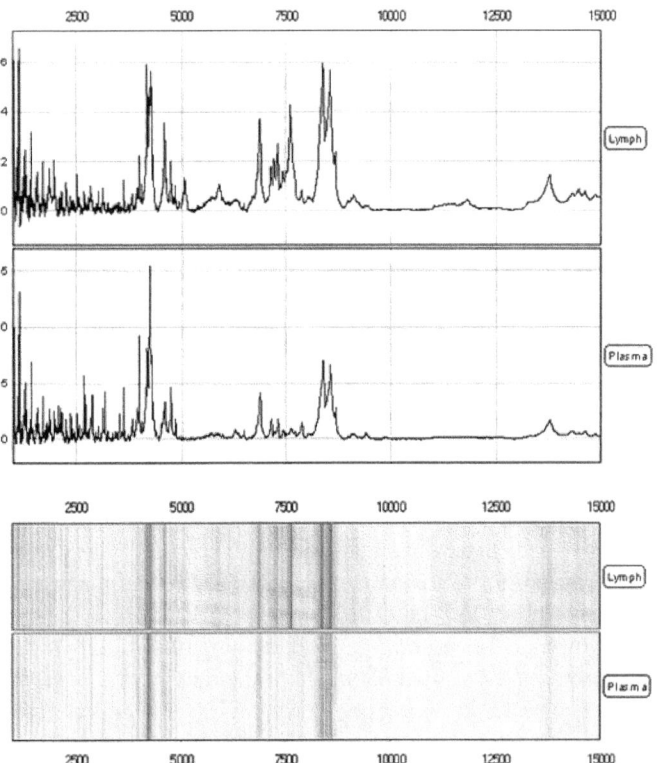

FIGURE 7. Surface-enhanced laser desorption ionization time-of-flight (SELDI-TOF) analysis of lymph and plasma from sheep during cardiopulmonary bypass. (**A**) The upper two mass chromatograms shown represent the SELDI proteomic profile from lymph and plasma, respectively, subsequent to cardiopulmonary bypass. The relative intensity is displayed along the y-axis and the mass is given as a molecular weight-to-charge ratio on the x-axis, and show low-molecular-weight ranges down to 250 Daltons. (**B**) The *lower two panels* show a gel view of the same mass chromatograms for lymph and plasma shown in panel **A**. The intensity of several peaks shown in the chromatograms appear as dark bands in the same relative position in panel **B**.

face with the laser beam. With SELDI-TO, hundreds of peptide and protein profiles from a small number of cells, or from a μL of a body fluid (such as serum, plasma, lymph, cerebrospinal fluid or urine) are easily obtained within minutes and displayed on the screen.[58] This technology has features that are complementary to 2-D PAGE analysis. However, unlike 2-D PAGE, this approach is able to profile proteins without regard to their intrinsic hydrophobicity, and it is very sensitive (attomole range) for the detection of proteins of lower molecular weights (<15 kDa), a range in which 2-D PAGE does not provide good resolution for detection. The Ciphergen's ProteinChip Systems are fully integrated platforms, comprising ProteinChip Arrays, instrumentation, and software. This system is designed for delivering fast and simple protein analysis using integrated platforms. The ProteinChip Biology System provides the researcher with a complete package that has the flexibility of conducting protein profiling and protein identification experiments using Ciphergen's ProteinChip Arrays.

To investigate the potential of SELDI-TOF technology for discovering diagnostic protein patterns of the lymph proteome, the lymph and plasma from normal sheep and from animals undergoing cardiopulmonary bypass were analyzed. The proteomic pattern generated from SELDI analysis of lymph and plasma showed that a characteristic and reproducible protein/peptide molecular weight profile was obtained, making it possible to differentiate normal lymph from that of plasma. In addition, a comparison analysis of normal lymph and plasma and samples from animals during cardiopulmonary bypass contained unique proteomic fingerprint patterns (FIG. 7). Thus, the application of proteomics to lymphatic research is indeed a very rich area for study, as it is very important to identify and characterize the proteomic content of lymph and the lymphatic endothelium, especially since the information content of the protein pattern might provide useful and important correlations with a specific biological state.

REFERENCES

1. WITTE, C.L. & M.H. WITTE. 1995. Disorders of lymph flow. Acad. Radiol **2**: 324–334.
2. WITTE, M.H., R. ERICKSON, M. BERNAS, et al. 1998. Phenotypic and genotypic heterogeneity in familial Milroy lymphedema. Lymphology **31**: 145.
3. ROCKSON, S.G. 2000. Lymphedema. Am. J. Med. **110**: 288–295.
4. SZUBA, A. & S.G. ROCKSON. 1997. Lymphedema: anatomy physiology and pathogenesis. Vasc. Med. **2**: 321–326.
5. KARKKAINEN, M.J., R.E FERRELL, E.C. LAWRENCE, et al. 2000. Missense mutations interfere with VEGFR-3 signaling in primary lymphoedema. Nat. Gen.. **25**: 153–159.

6. ENHOLM, B., T. KARPANEN, M. JELTSCH, *et al.* 2001. Adenoviral expression of vascular endothelial growth factor-C induces lymphangiogenesis in the skin. Circ. Res. **88:** 623–629.
7. JOUKOV, V., T. SORSA, V. JUMAR, *et al.* 1997. Proteolytic processing regulates receptor specificity and activity of VEGF-C. EMBO J. **16:** 3898-2911.
8. FITZ, L.J., J.C. MORRIS, P. TOWLER, *et al.* 1997. Characterization of murine Flt4 ligand/VEGF-C. Oncogene **15:** 613-618.
9. JELTSCH, M., A. KAIPAINEN, V. JOUKOV, *et al.* 1997. Hyperplasia of lymphatic vessels in VEGF-C transgenic mice. Science **275:** 1423–1425.
10. KARPANEN, T., M. EGEBALD, M.J. KARKKAUNEN, *et al.* 2001. Vascular endothelial growth factor C promotes tumor lymphangiogenesis and intralymphatic tumor growth. Cancer Res. **61:** 1786–1790.
11. KAIPAINEN, A., J. KORHONEN, T. MUSTONEN, *et al.* 1995. Expression of the Flt4 receptor tyrosine kinase becomes restricted to endothelium of lymphatic vessels and some high endothelial venules during development. Proc. Natl. Acad. Sci. USA **92:** 3566–3570.
12. VENTER, J.C., M.D. ADAMS, E.W. MEYER, *et al.* 2001. The sequence of the human genome Science **291:** 1304– 1351.
13. MCPHERSON, J.D., M. MARRA, L. HILLIER, *et al.* 2001. A physical map of the human genome. Nature **409:** 934–941.
14. WILKINS, M.R., E. GASTEIGER, J.C. SANCHEZ, *et al.* 1998. Two-dimensional gel electrophoresis for proteomic projects: the effects of protein hydrophobicity and copy number. Electrophoresis **19:** 1501–1505.
15. BLACKSTOCK, W.P. & M.P. WEIR. 1999. Proteomics: quantitative and physical mapping of cellular proteins. Trends Biotechnol. **17:** 121–127.
16. HANASH, S.M. 2000. Biomedical application of two dimensional electrophoresis using immobilized pH gradients:current status. Electrophoresis **21:** 1202–1209.
17. WILKINS, M.R., K.L. WILLIAMS, R.D. APPEL & D.F. HOCHSTRASSER, Eds. 1997. Research: New Frontiers in Functional Genomics. Springer-Verlag.
18. BEMIS, L.T. & P. SCHEDIN. 2000. Reproductive state of rat mammary gland stroma modulates human breast cancer migration and invasion. Cancer Res. **60:** 3414–3418.
19. MARGOLIS, L., S. HATFILL, R. CHUAQUI, *et al.* 1999. Long term organ culture of human prostate tissue in a NASA –designed rotating wall bioreactor. J. Urol. **161:** 290–297.
20. MCKEEHAN, W.L., P.S. ADAMS & D. FAST. 1987. Different hormonal requirements for androgen-independent growth of normal and tumor epithelial cells from rat prostate. In Vitro Cell Dev. Biol. **23:** 147–152.
21. CELIS, A., H.H. RAMSMUSSEN, P. CELIS, *et al.* 1999. Short-term culturing of low-grade superficial bladder transitional cell carcinomas leads to changes in the expression levels of several proteins involved in key cellular activities. Electrophoresis **20:** 355–561.
22. BONNER, R.F., M.R. EMMMERT-BUCK, K.. COLE, *et al.* 1997. Laser capture microdessiction : molecular analysis of tissue. Science **278:** 1481–183.
23. EMMERT-BUCK, M.R., R.F. BONNER, P.D. SMITH, *et al.* 1996. Laser capture microdissection. Science **274:** 998–1001.
24. ORNSTEIN, D., J.W. GILLESPIE, C.P. PAWELETZ, *et al.* 2000. Proteomic analysis of laser capture microdissected human prostate cancer and in vitro prostate cell lines. Electrophororesis **21:** 2235–2242.

25. EMMERT-BUCK, M., J.W. GILLLESPIE, C.P. PAWELETZ, et al. 2000. An approach to proteomic analysis of human tumors. Mol. Carcinogen. **27:** 158 165.
26. BANKS, R.E., M.J. DUNN, M.A. FORBES, et al. 1999. The potential use of laser capture microdissection to selectively obtain distinct populations of cells for proteomic analysis: preliminary findings. Electrophoresis **20:** 689–700.
27. ANDERSON, N.L. & N.G. ANDERSON. 1977. High resolution two-dimensional electrophoresis of human plasma proteins. Proc. Nat. Acad. Sci. USA **74:** 5421–5425.
28. GRIFFIN, T.J. & R. AEBERSOLD. 2001. Advances in proteomic analysis by mass spectrometry. J. Biol. Chem. **276:** 45497–45500.
29. O'FARRELL, P.H. 1975. High resolution two-dimensional electrophoresis of proteins, J. Biol. Chem. **250:** 4005–4021.
30. JUNG, E., M. HELLER, J-C. SANCHEZ & D.F. HOCHSTRASSER. 2000. Proteomics meets cell biology: the establishment of subcellular proteomes. Electrophoresis **21:** 3369–3377.
31. SHEVCHENKO, A., O.N. JENSEN, A.V. PODTELEJNIKOV, et al. 1996. Linking of genome and proteome by mass spectrometry: large scale identification of yeast proteins from two-dimensional gels. Proc. Nat. Acad. Sci. USA. **93:** 14440–14445.
32. KLOSE, J. 1975. Protein mapping by combined isoelectric focusing and electrophoresis of mouse tissues: a novel approach to testing for induced point mutations in mammals. Humangenetik **26:** 231–242.
33. HAYNES, P., I. MILLER, R. AEBERSOLD, et al. 1998. Proteins of rat serum: I Establishing a reference two-dimensional electrophoresis map by immunodetection and microbore high performance liquid chromatography-electrospray mass spectrometry. Electrophoresis **19:** 1484–1492.
34. WIILLIAMS, K.L. 1999. Genomes and proteomes: towards a multidimensional view of biology. Electrophoresis **20:** 678–688.
35. MANN, M., R.C. HENDRICKSON & A. PANDEY. 2001. Analysis of proteins and proteomes by mass spectrometry. Annu. Rev. Biochem. **70:** 437–473.
36. YATES, J.R., 3RD. 200? Mass spectrometry: from genomics to proteomics. Trends Genet. **16:** 5–8.
37. PAWELETZ, C.P., D.K.ORNSTEIN, M.J. ROTH, et al. 2000. Loss of annexin 1 correlates with early onset of tumorigenesis in esophageal and prostate carcinoma. Cancer Res. **60:** 6293–6297.
38. LIOTTA, L.A. & E.F. PETRICOIN. 2000. Molecular profiling of human cancer. Nature Rev. **1:** 48–56.
39. SANDISON, J.C. 1928. Observations on the growth of blood vessels as seen in the transparent chamber introduced into the rabbit's ear. Am. J. Anat. **41:** 475–496.
40. CLIFF, W.J. 1963. Observation on healing tissue: a combined light and electron microscope investigation. Phil. Trans. R. Soc. Lond. B. **246:** 305–325.
41. FOLKMAN, J. 1995. Clinical application of research on angiogenesis. N. Engl. J. Med. **222:** 1757–1763.
42. NICOSIA, R.F. 1987. Angiogenesis and the formation of lymphatic channels in cultures of thoracic ducts. In Vitro Cell Dev. Biol. **23:** 167–174.
43. LEAK, L.V. & M. JONES. 1994. Lymphangiogenesis in vitro: formation of lymphatic capillary-like channels from confluent minelayers of lymphatic endothelial cells. In Vitro Cell Dev. Biol. **30A:** 512–518.

44. PEPPER, M.S., S. WASI, N. FERRARA, et al. 1994. In vitro angiogenic proteolytic properties of bovine lymphatic endothelial cells. Exp. Cell Res. **210:** 298–305.
45. WAY, D.L., M.H. WITTE, M. FIALA, et al. 1993. Endothelial transdifferentiated phenotype and cell-cycle kinetics of AIDS-associated Kaposi's sarcoma cells. Lymphology **26:** 79–89.
46. LEAK, L.V., YU ZU-XI, M. JONES & V.J. FERRANS. 1999. Characterization of a transformed ovine lymphatic endothelial cell line. Microcirculation **6:** 63–73.
47. LEAK, L.V. & M. JONES. 1993. Lymphatic endothelium isolation characterization and long-term culture. Anat. Rec. **236:** 641–652.
48. LEAK, L.V., M. SAUNDERS, A.A. DAY & M. JONES. 2000. Stimulation of plasminogen activator and inhibitor in the lymphatic endothelium. Microvasc. Res. **60:** 201–211.
49. GORG, A., C. OBERMAIER, A. BOGUTH, et al. 2000. The current state of two-dimensional electrophoresis with immobilized pH gradients Electrophoresis **21:** 1037–1053.
50. GYGI, S.P. & R. AEBERSOLD. 2000. Mass spectrinetry and proteomics. Curr. Opin. Chem. Biol.. **4:** 489–494.
51. LIOTTA, L.A., EC. KOHN & E.F. PETRICOIN. 2001. Clinical proteomics, J. Amer. Med. Assoc. **286:** 2211–2214.
52. ARENKOV, P, A.. KUKHTIN, A. GEMMELL, et al. 2000. Protein microchips: use for immunoassay and enzymatic reactions. Anal. Biochem. **278:** 123–131.
53. GE, H. 2000. UPA, a universal protein array system for quantitative detection of protein-protein, protein-DNA, protein-RNA and protein-ligand interactions. Nucleic Acids Res. **28:** 3; MENDOZA, L.G. 1999. High throughput microarray based enzyme linked immunosorbant assay. BioTechnique **27:** 778–788.
54. PAWELETZ, C.P., L. CHARBONEAU, V.E. BICHSEL, et al. 2001. Reverse phase protein microarrays which capture disease progression show activation of pro-survival pathways at the cancer invasion front. Oncogene **20:** 1981–1989.
55. HUTCHENS, T.W. & T.T. YIP. 1993. New desorption strategies for the mass spectrometric analysis of macromolecules [rapid communication]. Mass. Spect. **7:** 576–580.
56. ARDEKANI, A.M., E.H. HERMAN, F.D. SISTARE, et al. 2001. Molecular profiling of cancer and drug-induced toxicity using new proteomic technology . Curr. Ther. Res. **62:** 803–819.
57. JOHNSTONE, R.A.W. & M.E. ROSE. 1996. Mass spectrometry for chemists and biochemists, 2nd ed. Cambridge University Press.
58. PAWELETZ, C.P., J.W. GILLESPIE, D.K. ORNSTEIN, et al. Rapid protein display profiling of cancer progression directly from human tissue using a protein biochip. Drug Dev. Res. 49:34–42, 2000.
59. SIMONE, N.L., C.P. PAWELETZ, L. CHARBONEAU, et al. 2000. Laser capture microdissection: beyond functional genomics to proteomics. Mol. Diagn. **5:** 301–307.
60. CHAMBERS, G., L. LAWRIE, P. CASH & G.I. MURRAY. 2000. Proteomics: a new approach to the study of disease. J. Pathol. **192:** 280–288.

Part 5: New Horizons
Panel Discussion

QUESTION: Is there a gap size over which the lymphatics will not eventually regenerate?

MELODY SWARTZ (*Northwestern University*): We haven't found that gap size yet. We technically can't make it any longer than 2 millimeters—otherwise we have problems with necrosis of the distal end of the tail. I think that the necrosis relates to restricting movement of the tail. At least a week is required before the tail becomes necrotic, so apparently this is not a problem with the blood supply.

Hydraulic Conductivity, Swelling, and Lymphatic Density

ALBERT MILLER (*Northwestern University*): The heart is an end organ, just as is the tail. All the lymph that goes up that tail starts at the tip, and all the lymph that leaves the heart starts in the heart muscle. The major lymphatic that leaves the heart is very easily visible after dye injection. We have removed approximately 5 to 6 millimeters of that vessel to see whether it regenerates. After 6 months, there is no evidence of regeneration in any of the dogs that were operated upon. We were able to see collaterals form, but no regeneration across that area.

SWARTZ: Tissues might have different regeneration rates, just as different tissues have different degrees of lymphatic density. Probably they differ in their requirements for fluid movement. In our mouse tail skin, the tissue swells quite readily, and the hydraulic conductivity is quite low. This creates the requirement for a nice, dense network of lymphatics. In contrast, one can examine a tissue like cartilage, where another mechanism to move fluid is present. It is like a sponge and, during physically activity, fluids are pushed in and out. Cartilaginous tissues don't need lymphatics at all. Therefore, the variables are likely to be the type of tissue and the time required for regeneration.

The relationship between reaction and diffusion of oxygen governs blood vessel density, so it is the relationship between hydraulic conductivity and ability to swell that governs the lymphatic density.

Relationship of Lymphangiogenesis to Gel Viscosity

QUESTION: If you reduced the viscosity of the gel, would you get more lymphangiogenesis through that region?

SWARTZ: We have experimented with three kinds of initial gel—I choose to call it initial gel; I don't like calling it a collagen gel. It is almost completely regenerated skin. It has hyaluronic acid, elastin fibers, and completely regenerated collagen. We only see lymphatic growth after 2 weeks, and by the first week, there is a re-epithelialization of that skin, so it almost doesn't matter what our starting material might be.

Flow as the Organizing Principle for Lymphangiogenesis

QUESTION: If you see blood vessels growing from both the left and the right side and, presumably there is an associated inflammatory reaction, with the release of growth factors from inflammatory cells, don't you think there may be some lymphangiostatic process occurring on the right side?

SWARTZ: We see flt-4-positive cells, but the organization seems to be from distal-to-proximal, the organization of fluid channels.

You raise several good points. With respect to the inflammatory response, we have stained, for example, for CD45-positive cells, which flood the matrix from day 1 to day 3. This is a component of the epithelial regeneration. Lymphangiogenesis, or, shall we say, the migration of LYVE-1-positive cells, occurs within a timeframe of 2 weeks. As early as day 5, we no longer see CD45-positive cells. But I do not want to give the impression that growth factors are not involved in this process. Surely, they are.

QUESTION: I'm inferring the opposite conclusion: that growth factors are not enough, because all of the growth factors for lymphatic and arterial cells would be present on the right side of the tail and the other cells won't migrate.

SWARTZ: This is where we would like to differentiate lymphangiogenesis from flow as the organizing principle. Flow is what determines the architecture. I believe that one can always make endothelial cells grow or sprout towards a source of growth factor, but to induce the cells to organize properly, one must induce the reason for them to live, to function.

The Patterning of the Superficial Lymphatics

COMMENT: There is obviously some exquisite patterning of lymphatics. It is clear that channels are forming, that the lymphatic vessels are lining those channels, but then they seem to end up in a very different organization, recapitulating this honeycomb network, presumably either against flow, or altering flow.

SWARTZ: If you have a circular vessel around an area, as opposed to two parallel pipes, the maximum distance for fluid to travel would be the radius. Therefore, one can envision these structures as networks of circles (which, when they grow together are really hexagons). These circles might represent the maximum distance that is physiologically optimal for fluid to travel.

The purpose of the superficial network is to drain fluid from the tissues. It is not to transport towards the tail. There are two deeper vessels, whose function is transport, so they don't require the hexagonal network. They carry the fluid that is collected by these circular structures.

Can Regeneration be Accelerated?

QUESTION: Can you accelerate regeneration?
SWARTZ: We are currently undertaking those investigations. We are making VEGF-C cross-linked fibrin gels to see how the network formation is altered after they are implanted.

Heterogeneity of Lymphatic Growth in Tumors

QUESTION: Perhaps these observations help to explain the heterogeneity of lymphatic vessel organization in various tumors in the presence of high levels of growth factors. For example, in different tumor types, when there are high levels of VEGF-C, one doesn't see uniform distribution: in some tumors, growth is very patchy, and in certain tumor types one observes an effect only at the periphery.

SWARTZ: It is not only a response to the growth factors that are released, but also to the fact that tumors are extremely proteolytic. In fact, those that have the highest rates of lymphatic metastasis are also the most proteolytically active along the edges.

Prevention of Protein Denaturation

QUESTION: How do you prevent protein denaturation during dissection and with the heat generated by laser application? And how do you make your monoclonal antibodies to each of these various proteins?

LANCE LIOTTA (*National Cancer Institute*): Laser-capture microdissection is a means to isolate cells from tissue. We have also used it to isolate components of sprouting endothelial cells in matrigel cultures: we can look at the central nidus of the sprout and the peripheral edge of the sprout. As an example, we can microdissect and analyze the proteomic or genomic fingerprint at those different levels.

Laser-capture microdissection does not appear to hurt the proteins or the DNA or RNA. That is because the laser does not directly hit the tissue. It hits a film that is in contact with the tissue. This then merges with the underlying tissue or fuses with it. The film contains a dye that absorbs all the laser energy. It is an extremely gentle technique. Essentially, the laser beam activates the film to stick in a gentle fashion to the underlying cells of interest; when the film is removed, only the cells that have been marked come away with the film.

We have shown by extensive studies that microdissection doesn't quantitatively or qualitatively affect the enzymatic activity of proteins or three-dimensional conformation, their post-translation modification, or anything else that we can measure.

We have also developed a number of extraction buffers and techniques to extract proteins from tissue cells in their native state, even if the tissue is stained. So if one uses the proper precipitating stains, the proteins will be preserved.

In laser desorption, the proteins are blasted off the surface of the bar, in the case of the cell-desurfacing machine. With the energy that is used to blast the proteins off, the whole protein flies; it is not fragmented in this technology. Other types of sophisticated mass spectroscopy actually break the proteins apart or ionize them.

Deriving Monoclonal Antibodies

QUESTION: How do you make your monoclonal antibodies to each of these various proteins?

LIOTTA: We have spent years qualifying our antibodies. We have 311 phospho-specific antibodies, and, additionally, 150 other types of antibodies in our repertoire. We have to validate those antibodies and ascertain that they have a single band on Western blotting, not just from purified proteins or even cultured cells, but from tissue itself.

Heuristic Systems

TIMOTHY PADERA (*Massachusetts Institute of Technology*): After your training sets, do your algorithms continue to learn from the data that you enter?

LIOTTA: The advantage of heuristic or learning systems, particularly the kind that we use, which feature both supervised and unsupervised learning, is that they continually learn and improve as more data are added to them. In fact, this system is designed to reside on the Web; the proteomic spectra can come from any part of the country or the world, the pattern can be compared to the ever-growing training set, and the system can learn, incorporating every patient that is diagnosed.

Every new patient that comes in is compared against every previous patient ever seen in the system. And, then, if that patient undergoes a qualification method (being developed with the assistance of the Food and Drug Administration) that validates the presence of a disease, the spectra of that patient can be added to the total spectra, contributing to the pattern used for future patients. We can thus incorporate each patient's data into an ever-growing database, and ultimately cull out patterns and clusters that correlate with disease.

Although we have a sensitivity and specificity of 95 and 98 percent, respectively, this is not sufficient for routine screening of patients for ovarian cancer. We must approximate 100 percent. We are striving to attain that elusive 100 percent sensitivity and specificity. We can add many, many patterns. They can be combinations of patterns or patterns of patterns, even if we don't know what those proteins are.

What Are These Proteins and Where Do They Come From?

QUESTION: Have you been able to analyze the identity of a specific peak? Is this the product of a cancer cell, or simply a cellular response to the cancer burden? How specific is the pattern to ovarian cancer? I think that you will probably have done some patterning with other cancers already.

LIOTTA: These are low molecular weight proteins. They probably are bound to other proteins because the kidney would otherwise clear them. They are probably cleavage fragments, enzymatic byproducts. And I think it would be a mistake to assume that they come from the cancer cell or even to try to find only those that come from the cancer cell.

It might be an elusive goal to look for unique cancer-specific markers. Cancer cells obviously are altered host-normal cells. The reason that they have their malignant behavior might reflect the relative abundance or activation of proteins, and not necessarily the presence of a unique protein that is only present in the cancer cell. The malignant behavior might also relate to the abnormal interaction with the host. Thereby, enzymes produced by the cancer cell or by the host cell or through any other host–tumor interactions could produce byproducts that can be detected in a proteomic pattern.

As an example, one can have a genetic defect, BRCA1, that produces breast cancer and ovarian cancer, but it doesn't produce cancers in other tissues. Obviously the genetic component in the tumor cell is not enough. It is the interaction within the unique host environment. Thus, we must look to that whole environment to produce the combinations of markers. The goal of a genetic algorithm is to find pattern markers that transcend the heterogeneity and have the ability to differentiate the population.

Is the Pattern for Ovarian Cancer Specific?

QUESTION: Is the pattern for ovarian cancer specific?

LIOTTA: The pattern appears to be specific for ovarian cancer in the limited studies we've done. We've looked at prostate cancer, which has a completely different pattern that has no similarity to the pattern of ovarian cancer. Furthermore, benign tumors, inflammatory conditions, other illnesses do not produce any pattern similar to that of the ovarian cancer. We have limited studies on other types of cancers, which also show different patterns.

The Classification Algorithm

QUESTION: Can you describe in more detail what the classification algorithm is and when we might expect to see it in print.

LIOTTA: Quite simply, we are looking at the relative amplitude of ion values in the spectrum and using the relative up-and-down values to form a point in n-dimensional space, where n is the number of different ion master charge values that we are choosing. The point in n-dimensional space is the relative amplitude, the Euclidian distance of those points. Essentially, it's a cluster analysis, where the points in the cluster are a pattern of the relative up-and-down heights of the peaks in the spectrum. We look at the pattern, and the clusters that it forms, in the normal versus the cancer cases. And the patterns in n-dimensional space form several clusters that are in the cancer patients and in the unaffected patients. Then, for an unknown patient, we look to see whether the pattern of those same subsets of proteins forms a point in space that falls within the existing normal clusters, within the existing tumor clusters or is separate.

Index of Contributors

Alitalo, K., 94–110
Ardekani, A.M., 211–228

Boardman, K.C., Jr., 197–210

Carmeliet, P., 80–93
Cassella, M., 120–130
Chaite, W., ix–x
Chang, L., 111–119

Ferrell, R.E., 39–51
Folkman, J., 111–119
Fusaro, V.A., 211–228

Gashev, A.A., 178–187

Harvey, N., 159–165
Hayward, A.R., 5–9

Jones, M., 211–228

Kaipainen, A., 111–119
Karkkainen, M.J., 94–110
Koh, J., 27–34

Lammie, P.J., 131–142

Leak, L.V., 211–228
Liotta, L.A., 211–228
Luttun, A., 80–93

Niklason, L.E., 27–34

Ochoa, E.R., 10–26
Oliver, G., 159–165
Olszewski, W.L., 52–63, 166–177

Paweletz, C.P., 211–228
Petricoin, E.F., III, 211–228
Punkosdy, G.A., 131–142

Rockson, S.G., 1–4, 64–75
Rosen, E.D., 143–158
Ross, S., 211–228

Saaristo, A., 94–110
Skobe, M., 120–130
Solan, A., 27–34
Swartz, M.A., 197–210

T.Cuenco, K., 131–142
Tjwa, M., 80–93

Vacanti, J.P., 10–26